Pituitary Today: Molecular, Physiological and Clinical Aspects

Frontiers of Hormone Research

Vol. 35

Series Editor

Ashley B. Grossman *London*

KARGER

Pituitary Today: Molecular, Physiological and Clinical Aspects

Volume Editors

Eduardo Arzt *Buenos Aires*
Marcello Bronstein *Sao Paulo*
Mirtha Guitelman *Buenos Aires*

22 figures, 1 in color, and 7 tables, 2006

KARGER

Basel · Freiburg · Paris · London · New York ·
Bangalore · Bangkok · Singapore · Tokyo · Sydney

Eduardo Arzt, PhD
Laboratorio de Fisiología y Biología Molecular
Departamento de Fisiología y Biología y Celular
FCEN – Universidad de Buenos Aires
Ciudad Universiaria. Pabellon II
Buenos Aires, Argentina

Mirtha Guitelman, MD
Endocrinology Division
Durand Hospital, and Neuroscience's Institute
University of Buenos Aires (INUBA)
Buenos Aires, Argentina

Marcello Bronstein, MD
Neuroendocrine Unit
Division of Endocrinology
and Metabolism
Hospital das Clinicas
University of Sao Paulo
Medical School, Sao Paulo, Brazil

Library of Congress Cataloging-in-Publication Data

Pituitary today: molecular, physiological and clinical aspects/volume
editors Eduardo Arzt, Marcello Bronstein, Mirtha Guitelman.
p.; cm. – (Frontiers of hormone research; v. 35)
Includes bibliographical references and index.
ISBN 3–8055–8155–6 (alk. paper)
1. Pituitary gland – Diseases. 2. Pituitary gland – Pathophysiology. I.

Bibliographic Indices. This publication is listed in bibliographic services, including Current Contents® and Index Medicus.

Disclaimer. The statements, options and data contained in this publication are solely those of the individual authors and contributors and not of the publisher and the editor(s). The appearance of advertisements in the book is not a warranty, endorsement, or approval of the products or services advertised or of their effectiveness, quality or safety. The publisher and the editor(s) disclaim responsibility for any injury to persons or property resulting from any ideas, methods, instructions or products referred to in the content or advertisements.

Drug Dosage. The authors and the publisher have exerted every effort to ensure that drug selection and dosage set forth in this text are in accord with current recommendations and practice at the time of publication. However, in view of ongoing research, changes in government regulations, and the constant flow of information relating to drug therapy and drug reactions, the reader is urged to check the package insert for each drug for any change in indications and dosage and for added warnings and precautions. This is particularly important when the recommended agent is a new and/or infrequently employed drug.

www.karger.com
Printed in Switzerland on acid-free paper by Reinhardt Druck, Basel
ISSN 0301–3073
ISBN-10: 3–8055–8155–6
ISBN-13: 978-3-8055-8155-4

Contents

Foreword

The pituitary and its disorders remain of perennial interest, and in a fast-developing field we make no apology for devoting another volume in this series to the subject, hard on the heels of *Molecular Pathology of the Pituitary* [vol. 32, 2004] Most intriguingly, this volume also developed from a meeting in South America, in this case one based at Iguazu Falls in Argentina, rather than Rio de Janeiro. Once again, a series of national and international experts converged on one of the most spectacular settings in the world, and I believe we have included in this volume many of the most scientifically exciting and clinically relevant areas in contemporary neuroendocrinology. But equally, I feel it also reveals much of the enthusiasm and dynamism of neuroendocrinology in South America, certainly in two of the largest countries Brazil and Argentina, but increasingly in many of the smaller South American countries. This is an exciting time in neuroendocrinology, and the South American continent is making salient and important contributions to its development.

Such meetings, held in surroundings of outstanding natural beauty, also encourage the initiation and deepening of many personal interactions, and I would like to underline my own deep thanks to the organizers and all my South-American colleagues. My daughter Sophie and I were subject to warm and welcoming local hospitality, and I will always fondly remember this conference. I would also like to take the opportunity to emphasize the totally unrestricted educational grant from Pfizer in supporting this meeting, demonstrating that international pharmaceutical companies can play a vital and, when required, disinterested role in the development of clinical science.

Ashley Grossman, North Devon

Preface

In November 1–4, 2005, in the spectacular setting of Iguazu falls, almost 140 professionals in medicine (endocrinologists, neurosurgeons), basic research and clinical biochemistry were hosted at the scientific meeting 'Pituitary Today', where all aspects related to the pituitary gland were discussed. Specialists from Latin America (including delegates from Argentina, Brazil, Mexico, Peru and Venezuela) gathered together for this event, together with expert speakers from Canada, USA, Germany, England and Australia who attended the event as invited guests.

Different aspects of pituitary physiopathology were covered in several symposia. Many of these lectures are now included in this book, including: Tumorigenesis (Melmed, Seilicovich, Arzt), Prolactinomas (Sarkar, Becu-Villalobos, Clapp), Hormone regulation and action (Drouin, Ghigo), GH/acromegaly (Ho, Bronstein, Goya), and HPA axis and Cushing (Grossman, De Nicola, Stalla). In addition, two consensus meetings on the diagnosis and treatment of pituitary disease (on acromegaly and prolactinomas) were discussed during the meeting, the one on prolactinomas being reported by Guitelman.

The meeting took place in a special environment where a scientific, cultural, and social exchange was predominant. In addition, we would like to emphasize that a group of young basic science and clinical fellows were also invited, presenting their results and thereby enriching the debates.

As coordinators of this event, we would like to express our thanks and recognition to Pfizer Laboratories who made 'Pituitary Today' possible in Latin

America; we also would like to mention the excellence of their staff whose working capacity allowed this meeting to attain an outstanding international level. We, as volume editors, thank all the contributors to this book of the series *Frontiers in Hormone Research* from Karger Publishers, Basel.

Mirtha Guitelman, Buenos Aires
Marcello D. Bronstein, Sao Paulo
Eduardo Arzt, Buenos Aires

Arzt E, Bronstein M, Guitelman M (eds): Pituitary Today: Molecular, Physiological and Clinical Aspects. Front Horm Res. Basel, Karger, 2006, vol 35, pp 1–8

Implication of Pituitary Tropic Status on Tumor Development

Ines Donangelo, Shlomo Melmed

Cedars-Sinai Research Institute, UCLA School of Medicine, Los Angeles, Calif., USA

Abstract

Pituitary tumor initiation and progression are associated with a plethora of genetic imbalances. Several genetic abnormalities have been described in pituitary tumors, from mutations in intracellular signaling (constitutive activation adenylyl cyclase) and growth factor pathways (epidermal growth factor receptor [EGFR]) to imbalance in cell cycle regulators (p16, p27, pRb). Unfortunately, most of these observations do not provide validated predictors of clinical behavior or of recurrence. The pituitary gland is notably plastic, and intrinsic and extrinsic stimuli result in profound growth changes ranging from hypoplasia to hyperplasia. The impact of pituitary tropic status on influencing neoplastic potential is difficult to test in human samples because the gland is not readily accessible for ongoing morphological observation. Animal models represent a functional approach to testing this hypothesis, and transgenic mouse models of pituitary tumor transforming gene (*PTTG*) inactivation or overexpression support the notion that pituitary tropic status directly correlates with likelihood for pituitary tumor formation. Understanding the mechanisms underlying changes in pituitary plasticity and their relationship to tumor development may contribute to the ability of regulating the development and progression of pituitary tumors.

Despite advances in understanding the pathogenesis of pituitary tumors, the primary etiology of pituitary tumorigenesis remains enigmatic. A wide range of genetic abnormalities has been found in pituitary tumors, from mutations in intracellular signaling pathways and expression of growth factors to imbalance in cell cycle regulators. However most of these observations do not provide validated predictors of clinical behavior or of recurrence.

The pituitary gland is notably plastic, and intrinsic and extrinsic stimuli result in profound growth changes ranging from hypoplasia to hyperplasia. Pituitary tropic status may directly correlate with the likelihood of tumor development;

however, this relationship has been shown to be difficult to prove in human specimens. This review will focus on discussing evidence for the relationship between pituitary cell tropic status and tumorigenesis.

Candidate Genes Involved in Pituitary Tumorigenesis

Pituitary adenomas are generally monoclonal in origin [1], suggesting that intrinsic molecular defects in a pituitary cell are likely the origin of a tumor. In addition to 'classical' hormone-secreting cells (somatotrope, lactotrope, mammosomatotrope, corticotrope, thyrotrope and gonadotrope), the pituitary also contains folliculostellate cells that secrete cytokines and growth factors [1, 2]. There is also increasing evidence for the existence of multifunctional [3] and progenitor cells in the adult pituitary gland [4, 5]. Abnormalities in several genes have been found in pituitary tumors (table 1); nevertheless, each genetic imbalance is noted only in a fraction of analyzed tumors. Moreover, pituitary tumors may arise from any of the pituitary cell subtypes, but in general specific genetic markers for different pituitary adenoma subtypes cannot be identified. An exception is the Gsα mutation found almost exclusively in GH-secreting adenomas [6–9]. The resulting oncogene *gsp* caused constitutively activates adenylyl cyclase with resulting cell growth stimulation. Despite being a specific marker for somatotropinomas, the *gsp* oncogene is present in only 30–50% of the cases [10–12], mostly in smaller tumors [7] that are more responsive to somatostatin analogs [13].

Pituitary Tropic Status as a Modulator of Tumor Formation

Under physiological conditions, hypothalamic and peripheral hormones are the major contributing factor for pituitary tropic activity. The effects of age (puberty) and pregnancy/lactation on increasing pituitary volume are widely known, and prolonged hypothyroidism and estrogen excess are recognized causes of pituitary hyperplasia [14]. However, there is no direct evidence that pituitary hyperplasia is a sufficient major cause of pituitary tumor development. Hyperplasic proliferation of prolactin-secreting cells during pregnancy and lactation does not increase the frequency of prolactinomas [15], and untreated primary hypothyroidism and exogenous estrogen administration are infrequently associated with the development of adenomas [16, 17]. Moreover, although GH-secreting adenomas develop in patients with ectopic tumors secreting growth hormone-releasing hormone (GHRH), most of these patients usually develop acromegaly solely due to somatotrope hyperplasia. These observational reports in humans contrast with findings in animal models for

Table 1. Selected genes involved in pituitary tumor development

Gene	Mechanism of activation or inactivation	Defect	References
Activating			
gsp	point mutation	activation of cAMP pathway	[8–12]
CREB	Increased Ser-phosphorylated CREB promoted by gsp overexpression	dimerizes with cAMP response elements	[29, 30]
Cyclin D1	allelic imbalance	cell cycle progression	[31]
HMGA2	amplification	rearrangement of chromatin structure?	[32]
EGFR	unknown	EGF pathway activates cell growth	[33, 34]
PTTG	unknown	disrupted cell cycle, chromosomal instability, transactivation activity	[35–37]
Inactivating			
RB1	gene silencing through methylation	loss of pRb → unrestrained cell cycle progression	[38, 39]
CDKN2A	Gene silencing through methylation	loss of p16→ unrestrained cell cycle progression	[40, 41]
CIP1/KIP1	Unknown in humans (methylation in rodent cells lines)	loss of p27→ unrestrained cell cycle progression	[42–44]

CREB encodes cAMP-responsive element binding protein. *HMGA2* = High Mobility Group A2 gene; EGFR = epidermal growth factor receptor; *RB1* = retinoblastoma tumor suppressor gene; CIP1/KIP encodes p27; CDKN2A encodes p16.

Adapted with modifications from Melmed S, Kleineberg D: Anterior pituitary; in Larsen PR, Kronenberg HM, Melmed S, Polonsky KS (eds): Williams Textbook of Endocrinology, ed 10. Saunders, Philadelphia, 2003, pp 177–280.

pituitary hyperplasia. Pituitary hyperplasia in GHRH transgenic mice [18] or due to the exogenous administration of estrogen [19] ultimately results in tumor formation. The reason for the species differences is not clear, but prolonged sustained hyperplasia maybe is required for tumor development.

Although pituitary tropic stimuli do not frequently originate tumors in humans, they may influence the intrapituitary milieu to either enhance or attenuate the expansion of a monoclonal tumor cell population. In mouse models with combined hypoplasic and tumorigenic genetic abnormalities, tumor incidence is lower than in mice harboring the tumorigenic genetic change alone. This pattern is observed in both compound *Pttg−/−Rb+/−* mice (inactivation of *pttg* leads to pituitary hypoplasia, and *Rb* disruption causes pituitary tumor) and in double-mutant *p18Ink4c−/−cdk4−/−* mice (p18 inactivation causes pituitary hyperplasia that progresses to tumor, and loss of cdk4 results in pituitary hypoplasia) [20, 21]. Understanding the mechanisms that regulate

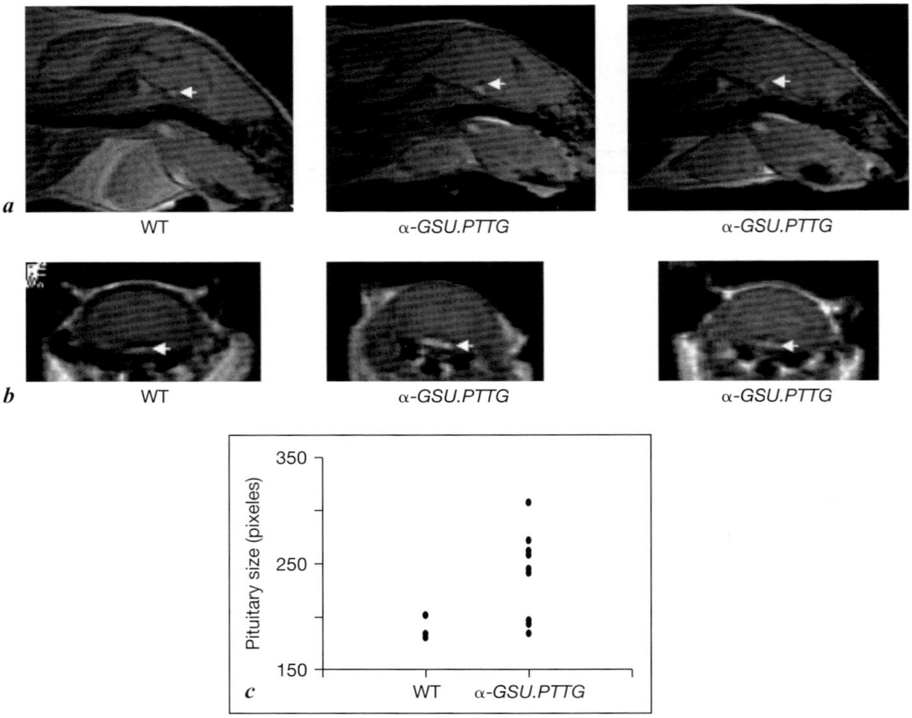

Fig. 1. Evidence of pituitary enlargement on magnetic resonance images (MRI) in *PTTG* transgenic mice. Sagittal (***a***) and coronal (***b***) MRI images of one wild-type (WT) and two transgenic (α-*GSU.PTTG*) mice. ***c*** Scattergram depicting pituitary size in total pixels obtained by adding the pituitary area obtained from consecutive sagittal images. Reproduced from Abbud et al. [26], with permission.

pituitary plasticity therefore provides insights for disrupting the development and progression of pituitary tumors.

PTTG Transgenic Mice: Models for the Pituitary Tropic Status and Tumorigenesis Correlation

Pituitary tumor transforming gene (*PTTG)* was isolated from a pituitary tumor cell line and its overexpression results in cellular transformation in vitro and tumor formation in nude mice [22]. *PTTG* was identified as the index mammalian securin, regulating sister chromatid separation during mitosis [23], and excessive or suppressed *PTTG* levels result in aneuploidy [24, 25]. *PTTG* abundance correlates with pituitary gland tropic status. Transgenic mouse models of

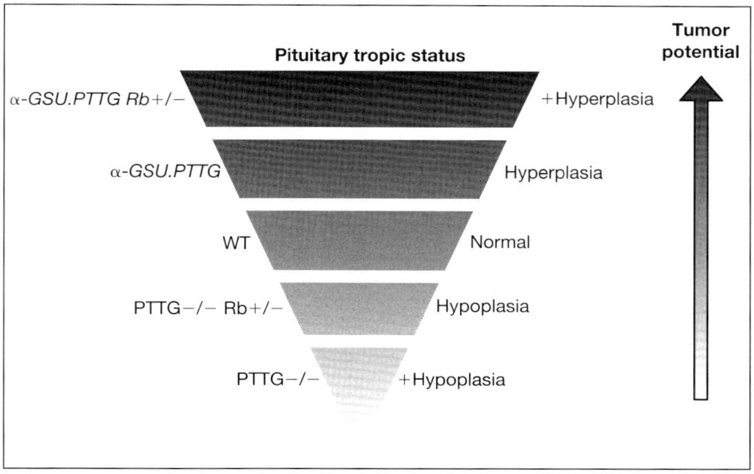

Fig. 2. Pituitary PTTG content correlates with gland tropic status and tumor formation potential. On the left side of the inverted triangle mouse models with descending pituitary PTTG content, with or without the combination with tumorigenic Rb+/−, are listed. The right side of the inverted triangle lists the observed effects of the different genotypes on pituitary tropic status and correlation with pituitary tumorigenic potential (arrow).

both *PTTG* overexpression and inactivation support an enabling role for changes in *PTTG* levels on development of pituitary hyperplasia and hypoplasia, and on conferring tumor growth advantage or tumor growth protection, respectively. Mice with transgenic human *PTTG1* expression driven by the α-subunit glycoprotein (α-GSU) promoter express *PTTG* in LH-, FSH-, TSH- and GH-secreting cells [26]. α-*GSU.PTTG* mice also develop frequent plurihormonal hyperplasia (fig. 1). Increased serum LH, testosterone, GH and IGF-I levels result in marked seminal vesicle and prostate enlargement in male α-*GSU.PTTG* mice. In contrast, *Pttg* inactivation results in tissue-specific opposing tropic effects, i.e. pituitary, pancreatic β-cell, splenic and testicular hypoplasia [25].

As *PTTG* abundance correlates with pituitary gland tropic status, *PTTG* transgenic mouse models may subserve the study of pituitary plasticity and tumorigenesis association. Mice with heterozygous inactivation of retinoblastoma gene (*Rb+/–*) represent a well established model for pituitary tumors [27, 28]. *Pttg* inactivation in *Rb+/−* mice results in relative protection from tumor development [20]. While *Rb+/−* mice have a cumulative pituitary tumor incidence of 86% by 13 months, only 30% of compound *Pttg−/−Rb+/−* mice develop pituitary tumors by the same age ($p < 0.01$) [20]. In contrast, combined *Rb+/−* and targeted pituitary *PTTG* over-expression enhances pituitary hyperplasia and tumor prevalence. Compound α-*GSU.PTTGxRb+/−* mice have enlarged pituitary glands, and 3.5-fold increase in the frequency of tumors originating from α-subunit expressing cells [submitted]. A proposed depiction of effects resulting from in vivo changes in pituitary *PTTG* content is shown in figure 2.

Conclusion

The molecular pathogenesis of pituitary tumors involves diverse mechanisms, and progress in the search for primary causal abnormalities has been modest. More likely, pituitary tumors develop as a result of a cascade of permissive abnormal events, in which the tropic status may represent either an obstacle or a permissive growth milieu. Understanding the mechanisms for controlling pituitary plasticity may therefore result in the ability to regulate development and progression of pituitary tumorigenesis.

References

1 Clayton RN, Farrell WE: Pituitary tumour clonality revisited. Front Horm Res. Basel, Karger, 2004, vol 32, pp 186–204.
2 Allaerts W, Vankelecom H: History and perspectives of pituitary folliculo-stellate cell research. Eur J Endocrinol 2005;153:1–12.
3 Villalobos C, Nunez L, Garcia-Sancho J: Anterior pituitary thyrotropes are multifunctional cells. Am J Physiol 2004;287:E1166–E1170.
4 Krylyshkina O, Chen J, Mebis L, Denef C, Vankelecom H: Nestin-immunoreactive cells in rat pituitary are neither hormonal nor typical folliculo-stellate cells. Endocrinology 2005;146: 376–2387.
5 Chen J, Hersmus N, Van DV, Caesens P, Denef C, Vankelecom H: The adult pituitary contains a cell population displaying stem/progenitor cell and early embryonic characteristics. Endocrinology 2005;146:3985–3998.
6 Kim HJ, Kim MS, Park YJ, Kim SW, Park DJ, Park KS, Kim SY, Cho BY, Lee HK, Jung HW, Han DH, Lee HS, Chi JG: Prevalence of Gs alpha mutations in Korean patients with pituitary adenomas. J Endocrinol 2001;168:221–226.
7 Landis CA, Harsh G, Lyons J, Davis RL, McCormick F, Bourne HR: Clinical characteristics of acromegalic patients whose pituitary tumors contain mutant Gs protein. J Clin Endocrinol Metab 1990;71:1416–1420.
8 Spada A, Arosio M, Bochicchio D, Bazzoni N, Vallar L, Bassetti M, Faglia G: Clinical, biochemical, and morphological correlates in patients bearing growth hormone-secreting pituitary tumors with or without constitutively active adenylyl cyclase. J Clin Endocrinol Metab 1990;71:1421–1426.
9 Spada A, Vallar L: G-protein oncogenes in acromegaly. Horm Res 1992;38:90–93.
10 Johnson MC, Codner E, Eggers M, Mosso L, Rodriguez JA, Cassorla F: Gps mutations in Chilean patients harboring growth hormone-secreting pituitary tumors. J Pediatr Endocrinol Metab 1999;12:381–387.
11 Yang I, Park S, Ryu M, Woo J, Kim S, Kim J, Kim Y, Choi Y: Characteristics of gsp-positive growth hormone-secreting pituitary tumors in Korean acromegalic patients. Eur J Endocrinol 1996;134:720–726.
12 Yasufuku-Takano J, Takano K, Morita K, Takakura K, Teramoto A, Fujita T: Does the prevalence of gsp mutations in GH-secreting pituitary adenomas differ geographically or racially? Prevalence of gsp mutations in Japanese patients revisited. Clin Endocrinol (Oxf) 2006;64: 91–96.

13 Faglia G, Spada A: Genesis of pituitary adenomas: state of the art. J Neurooncol 2001;54:95–110.
14 Horvath E, Kovacs K, Scheithauer BW: Pituitary hyperplasia. Pituitary 1999;1:169–179.
15 Coogan PF, Baron JA, Lambe M: Parity and pituitary adenoma risk. J Natl Cancer Inst 1995;87: 1410–1411.
16 Ghannam NN, Hammami MM, Muttair Z, Bakheet SM: Primary hypothyroidism-associated TSH-secreting pituitary adenoma/hyperplasia presenting as a bleeding nasal mass and extremely elevated TSH level. J Endocrinol Invest 1999;22:419–423.
17 Kovacs K, Stefaneanu L, Ezzat S, Smyth HS: Prolactin-producing pituitary adenoma in a male-to-female transsexual patient with protracted estrogen administration: a morphologic study. Arch Pathol Lab Med 1994;118:562–565.
18 Asa SL, Kovacs K, Stefaneanu L, Horvath E, Billestrup N, Gonzalez-Manchon C, Vale W: Pituitary adenomas in mice transgenic for growth hormone-releasing hormone. Endocrinology 1992;131:2083–2089.
19 Heaney AP, Horwitz GA, Wang Z, Singson R, Melmed S: Early involvement of estrogen-induced pituitary tumor transforming gene and fibroblast growth factor expression in prolactinoma pathogenesis. Nat Med 1999;5:1317–1321.
20 Chesnokova V, Kovacs K, Castro AV, Zonis S, Melmed S: Pituitary hypoplasia in Pttg−/− mice is protective for Rb+/− pituitary tumorigenesis. Mol Endocrinol 2005;19:2371–2379.
21 Pei XH, Bai F, Tsutsui T, Kiyokawa H, Xiong Y: Genetic evidence for functional dependency of p18Ink4c on Cdk4. Mol Cell Biol 2004;24:6653–6664.
22 Pei L, Melmed S: Isolation and characterization of a pituitary tumor-transforming gene (PTTG). Mol Endocrinol 1997;11:433–441.
23 Zou H, McGarry TJ, Bernal T, Kirschner MW: Identification of a vertebrate sister-chromatid separation inhibitor involved in transformation and tumorigenesis. Science 1999;285:418–422.
24 Yu R, Lu W, Chen J, McCabe CJ, Melmed S: Overexpressed pituitary tumor-transforming gene causes aneuploidy in live human cells. Endocrinology 2003;144:4991–4998.
25 Wang Z, Yu R, Melmed S: Mice lacking pituitary tumor transforming gene show testicular and splenic hypoplasia, thymic hyperplasia, thrombocytopenia, aberrant cell cycle progression, and premature centromere division. Mol Endocrinol 2001;15:1870–1879.
26 Abbud RA, Takumi I, Barker EM, Ren SG, Chen DY, Wawrowsky K, Melmed S: Early multipotential pituitary focal hyperplasia in the alpha-subunit of glycoprotein hormone-driven pituitary tumor-transforming gene transgenic mice. Mol Endocrinol 2005;19:1383–1391.
27 Jacks T, Fazeli A, Schmitt EM, Bronson RT, Goodell MA, Weinberg RA: Effects of an Rb mutation in the mouse. Nature 1992;359:295–300.
28 Leung SW, Wloga EH, Castro AF, Nguyen T, Bronson RT, Yamasaki L: A dynamic switch in Rb+/− mediated neuroendocrine tumorigenesis. Oncogene 2004;23:3296–3307.
29 Bertherat J, Chanson P, Montminy M: The cyclic adenosine 3′,5′-monophosphate-responsive factor CREB is constitutively activated in human somatotroph adenomas. Mol Endocrinol 1995;9:777–783.
30 Peri A, Krausz C, Cioppi F, Granchi S, Forti G, Francavilla S, Serio M: Cyclic adenosine 3′,5′-monophosphate-responsive element modulator gene expression in germ cells of normo- and oligoazoospermic men. J Clin Endocrinol Metab 1998;83:3722–3726.
31 Hibberts NA, Simpson DJ, Bicknell JE, Broome JC, Hoban PR, Clayton RN, Farrell WE: Analysis of cyclin D1 (CCND1) allelic imbalance and overexpression in sporadic human pituitary tumors. Clin Cancer Res 1999;5:2133–2139.
32 Finelli P, Pierantoni GM, Giardino D, Losa M, Rodeschini O, Fedele M, Valtorta E, Mortini P, Croce CM, Larizza L, Fusco A: The High Mobility Group A2 gene is amplified and overexpressed in human prolactinomas. Cancer Res 2002;62:2398–2405.
33 Chaidarun SS, Eggo MC, Sheppard MC, Stewart PM: Expression of epidermal growth factor (EGF), its receptor, and related oncoprotein (erbB-2) in human pituitary tumors and response to EGF in vitro. Endocrinology 1994;135:2012–2021.
34 Onguru O, Scheithauer BW, Kovacs K, Vidal S, Jin L, Zhang S, Ruebel KH, Lloyd RV: Analysis of epidermal growth factor receptor and activated epidermal growth factor receptor expression in pituitary adenomas and carcinomas. Mod Pathol 2004;17:772–780.

35 Zhang X, Horwitz GA, Prezant TR, Valentini A, Nakashima M, Bronstein MD, Melmed S: Structure, expression, and function of human pituitary tumor-transforming gene (PTTG). Mol Endocrinol 1999;13:156–166.
36 Zhang X, Horwitz GA, Heaney AP, Nakashima M, Prezant TR, Bronstein MD, Melmed S: Pituitary tumor transforming gene (PTTG) expression in pituitary adenomas. J Clin Endocrinol Metab 1999;84:761–767.
37 Pei L: Identification of c-myc as a down-stream target for pituitary tumor-transforming gene. J Biol Chem 2001;276:8484–8491.
38 Donangelo I, Marcos HP, Araujo PB, Marcondes J, Filho PN, Gadelha M, Chimelli L: Expression of retinoblastoma protein in human growth hormone-secreting pituitary adenomas. Endocr Pathol 2005;16:53–62.
39 Simpson DJ, Hibberts NA, McNicol AM, Clayton RN, Farrell WE: Loss of pRb expression in pituitary adenomas is associated with methylation of the RB1 CpG island. Cancer Res 2000;60: 1211–1216.
40 Woloschak M, Yu A, Post KD: Frequent inactivation of the p16 gene in human pituitary tumors by gene methylation. Mol Carcinog 1997;19:221–224.
41 Simpson DJ, Bicknell JE, McNicol AM, Clayton RN, Farrell WE: Hypermethylation of the p16/CDKN2A/MTSI gene and loss of protein expression is associated with nonfunctional pituitary adenomas but not somatotrophinomas. Genes Chromosomes Cancer 1999;24:328–336.
42 Qian X, Jin L, Kulig E, Lloyd RV: DNA methylation regulates p27kip1 expression in rodent pituitary cell lines. Am J Pathol 1998;153:1475–1482.
43 Lloyd RV, Jin L, Qian X, Kulig E: Aberrant p27kip1 expression in endocrine and other tumors. Am J Pathol 1997;150:401–407.
44 Jin L, Qian X, Kulig E, Sanno N, Scheithauer BW, Kovacs K, Young WF, Jr, Lloyd RV: Transforming growth factor-beta, transforming growth factor-beta receptor II, and p27Kip1 expression in nontumorous and neoplastic human pituitaries. Am J Pathol 1997;151:509–519.

Shlomo Melmed, MD
Cedars-Sinai Medical Center
8700 Beverly Blvd., Room 2015
Los Angeles, CA 90048 (USA)
Tel. +1 310 423 4691, Fax +1 310 423 0119, E-Mail melmeds@cshs.org

Arzt E, Bronstein M, Guitelman M (eds): Pituitary Today: Molecular, Physiological and Clinical Aspects. Front Horm Res. Basel, Karger, 2006, vol 35, pp 9–21

Anterior Pituitary Cell Renewal during the Estrous Cycle

M. Candolfi[b], *V. Zaldivar*[a], *G. Jaita*[a], *A. Seilicovich*[a]

[a]Centro de Investigaciones en Reproducción, Facultad de Medicina, Universidad de Buenos Aires, Buenos Aires, Argentina; [b]Gene Therapeutics Research Institute, Cedars Sinai Medical Center and David Geffen School of Medicine, UCLA, Los Angeles, Calif., USA

Abstract

The anterior pituitary gland undergoes a process of cell renewal during the estrous cycle. Although the occurrence of proliferation and death of anterior pituitary cells at specific stages of the estrous cycle is well known, the underlying mechanisms that regulate these processes are still being uncovered. In spite of the recognized proliferative effects of estrogens on lactotropes, recent evidence shows that estrogens can also trigger antiproliferative and apoptotic responses in anterior pituitary cells. In the present review we analyze the actions of gonadal steroids on proliferation and death of anterior pituitary cells during the estrous cycle and the mediators involved in these actions. Estradiol sensitizes anterior pituitary cells not only to mitogenic stimuli but also to apoptotic signals and upregulates local synthesis of tropic growth factors as well as proapoptotic cytokines. Several growth factors and cytokines have been shown to induce estrogen-dependent lactotrope proliferation and death, whereas progesterone antagonizes estrogen-induced effects. These locally synthesized factors may mediate the effects of gonadal steroids in the process of anterior pituitary cell renewal during the estrous cycle.

Anterior pituitary functions, including hormone secretion and maintenance of tissue homeostasis, depend on hormones and neurotransmitters released from the hypothalamus and the neural lobe, and on systemic hormones secreted in the periphery, mainly from target glands. Also, hormones, growth factors and cytokines synthesized locally in the pituitary gland regulate its function in an autocrine, juxtacrine and/or paracrine manner.

Although the anterior pituitary gland was classically considered a quiescent organ, in recent years this concept was reviewed and the plasticity of this gland is now well recognized [1, 2]. Appropriate stimuli induce mitosis, apoptosis and cell differentiation during physiological conditions such as pregnancy, lactation and the estrous cycle. At the end of pregnancy and during lactation, the lactotrope subpopulation undergoes proliferation induced by high estrogen levels and neuroendocrine responses triggered by suckling [3]. The bidirectional interconversion between lactotropes and somatotropes has also been implicated in the expansion of the lactotrope subpopulation during lactation and the reduction in the number of lactotropes at the end of this period [4]. However, strong evidence supports the idea that cell death is an important mechanism to reduce the size of the lactotrope subpopulation to values similar to those observed before pregnancy [5]. In fact, cell proliferation and death at specific stages of the estrous cycle have been implicated in cell renewal of the anterior pituitary gland during each estrous cycle, though the underlying mechanisms are in the process of being discovered.

Anterior Pituitary Cell Renewal

In the anterior pituitary gland, about 1.5% of the cells are renewed daily [6]. Lactotropes are the cell type with the highest rate of proliferation in this gland, followed by somatotropes, corticotropes and gonadotropes [7]. Anterior pituitary cells show a circadian rhythm in proliferation in both males and females [7, 8]. However, lactotrope proliferation rates, higher in females than in males [7], are reduced by ovariectomy [1, 9], indicating that gonadal steroids are involved in anterior pituitary cell turnover. In fact, the anterior pituitary gland undergoes a process of cell renewal during the estrous cycle in the female rat that seems to be dependent on cyclic changes in circulating sex hormone levels [7, 10, 11]. Plasma levels of estradiol begin to rise at diestrus, reach peak values at mid-proestrus and then rapidly fall to basal levels on the morning of estrus, while systemic levels of progesterone show a peak on the first day of diestrus and a higher one in the evening of proestrus [12]. This hormonal profile has been suggested to modulate both cell proliferation and death in the anterior pituitary gland during the estrous cycle [13]. Pituitary cell proliferation takes place at estrus whereas apoptosis occurs at proestrus, the lactotropes being the anterior pituitary cell subpopulation with the highest turnover [8, 10, 14].

Anterior Pituitary Cell Proliferation

In estrogen-responsive tissues such as mammary gland, uterus and anterior pituitary gland, mainly in the lactotrope subpopulation, estrogens act as potent mitogens [1, 2]. However, increasing evidence suggests that the mitogenic action of estrogens on lactotropes is not a direct effect but depends on the secretion of both hypothalamic and local factors. In fact, the delay between uterine endometrial cell [15] and lactotrope proliferation after the proestrus peak of estrogens supports an indirect effect of estrogens on the pituitary gland. The estrogenic action on lactotropes may be partially exerted through its inhibitory effect on the hypothalamic release of dopamine, a potent inhibitor of lactotrope proliferation. Dopamine, through its D2 receptor short isoform reduces lactotrope proliferation by inducing the expression of an autocrine antiproliferative growth factor, transforming growth factor-β1 (TGF-β1), and its receptor TGF-β1 type II [16]. The hypothalamic release of LHRH triggered by estradiol in the afternoon of proestrus may also be involved in lactotrope proliferation at estrus, this action of LHRH probably being mediated through angiotensin II released from gonadotropes [17]. However, a body of evidence shows that mitogenic actions of estradiol in the anterior pituitary gland are mainly exerted by increasing the production of local growth factors, which in turn promote proliferation and differentiation of lactotropes [1, 2]. Also, estradiol may expand the lactotrope subpopulation by restraining the local synthesis of antiproliferative autocrine and/or paracrine factors [1, 2].

Several growth factors, in particular, epidermal growth factor (EGF), TGF-α, TGF-β, basic fibroblast growth factor (bFGF), insulin-like growth factor-I (IGF-I) and IGF-II, among others, have shown to be involved in lactotrope cell growth [1, 2, 18]. TGF-α, a recognized local mediator of the mitogenic effect of estrogens, is produced in lactotropes and corticotropes, and shows mitogenic activity in both cell types. Estradiol increases the expression of TGF-α and its homolog EGF as well as EGF receptors in the anterior pituitary. The synthesis of TGF-β3 from lactotropes is also stimulated by estradiol and mediates, in part, the mitogenic effect of estradiol. TGF-β3 may not act in an autocrine manner on lactotrope proliferation, but rather through the paracrine action of bFGF synthesized by folliculostellate cells [1, 2]. Chronic administration of estradiol also increases the anterior pituitary expression of IGF-I [19]. It has been proposed that estradiol and growth hormone (GH) increase IGF-I release, which in turn may stimulate lactotrope proliferation through vasoactive intestinal peptide (VIP) [20]. VIP increases nerve growth factor (NGF) synthesis, which besides stimulating lactotrope proliferation would act as an autocrine mediator of the transdifferentiation of somatolactotropes to lactotropes [1]. Galanin, a neuropeptide produced by lactotropes in female rats and whose synthesis is stimulated by estradiol, appears to be involved in the stimulatory effect of estrogens on lactotrope proliferation [2]. On the contrary, estradiol decreases the expression of TGF-β1, an inhibitor of lactotrope proliferation [1, 2].

Although it is recognized that estrogens exert mitogenic actions on the lactotrope subpopulation, the effect of estrogens in the generation of prolactinomas is a subject of controversy [21]. In fact, the induction of prolactinomas in rodents after estrogen administration is dependent on the strain: while Fischer 344 rats are extremely sensitive to estradiol or diethylbestrol treatment, other strains, such as Sprague-Dawley, are resistant to generate pituitary adenomas [22, 23]. Although the molecular basis of these differences have not been completely elucidated, the expression of estrogen receptors is similar in Fischer 344 and Sprague-Dawley strains [24], which suggests that downstream pathways underlie the differences in sensitivity to tumor promotion between strains. In fact, the expression of some tumor suppressor proteins is decreased or absent in the pituitary of Fischer 344 rats. Loss of pRb, a tumor suppressor protein that regulates G1 to S transition of the cell cycle, has been related to tumor susceptibility in Fischer 344 rats to diethylstilbestrol [25]. Moreover, p53, another tumor suppressor protein, remains undetectable in the pituitary of Fischer 344 rats after estrogen treatment whereas in Sprague-Dawley rats p53 and pRb are even enhanced by this steroid [26]. Although p53 does not appear to undergo mutations in human pituitary adenomas [21], since p53 is a key factor in the surveillance of DNA stability, and induces apoptosis when DNA cannot be repaired, the inhibition of p53-induced apoptosis could promote tumor progression. Also, a gene within the chromosomal interval of the estrogen-dependent pituitary mass on chromosome 5 (Edpm5) has been reported to regulate the switch to an angiogenic phenotype in Fischer 344 rat tumors [26]. On the other hand, folliculostellate cells from Fischer 344 rats but not from Sprague Dawley rats increase the mitogenic response of lactotropes to estradiol and produce higher levels of estrogen-induced bFGF suggesting that the interactions between folliculostellate cells and lactotropes may also be involved in the differential susceptibility in the transition of hyperplasia to prolactinoma between rat strains [27].

Estrogens can also exert antiproliferative effects on anterior pituitary cells. It has been reported that chronic estrogen treatment inhibits the growth of the MtT/W15 tumor, reducing the number of lactotropes [28]. Also, estradiol inhibits in a dose-dependent manner the proliferation of the somatolactotrope cell line GH3 [29]. Although a dominant negative estrogen receptor-α (ERα) induces apoptosis in GH4 lactotrope cells, overexpression of wild-type ERα inhibits estrogen-induced proliferation in this cell line [30]. Moreover, since short-term estradiol treatment reduces the mitogenic effect of insulin and IGF-I in primary cultures of pituitary cells, it was suggested that the net effect of estradiol on lactotrope proliferation reflects two opposite actions: a short latency antimitogenic action occurring in the presence of growth factors, and a long-term mitogenic action that requires the synthesis of local mediators [31, 32]. Furthermore, estradiol stimulates the local release of tumor necrosis factor-α (TNF-α) and sensitizes anterior pituitary cells to the antiproliferative effect of this cytokine [33, 34]. Estradiol also stimulates the release of interleukin 6 (IL-6), a member of the gp130 cytokine family, from lactotrope and somatolactotrope tumoral cells [35]. IL-6 has paradoxical effects on anterior pituitary cell proliferation. This cytokine is produced by folliculostellate cells in the normal anterior pituitary, whereas in the majority of pituitary adenomas, in which folliculostellate cells are rare or absent, tumor cells are the source of IL-6. IL-6 inhibits the growth of normal pituitary cells but stimulates pituitary tumor cell proliferation [18]. Although the underlying mechanism for these opposing growth effects of IL-6 on normal and adenomatous pituitary cells has not been elucidated yet, it has been suggested that differences in the signaling pathways may be responsible for these different mitogenic effects [35].

Apoptosis in the Anterior Pituitary Gland

During the estrous cycle, the apoptotic rate of the pituitary gland peaks in the afternoon of proestrus [14]. In basal conditions, apoptosis has been observed in gonadotropes, corticotropes, lactotropes and somatotropes in the anterior pituitary gland [6, 11].

One of the cytokine receptor families expressed in the anterior pituitary gland is the TNF superfamily whose members have been reported to initiate signaling cascades leading to cell death. TNF receptors (TNFR) and Fas that belong to this cytokine receptor family are critical regulators of the balance between two opposing processes: binding to their cognate receptors can lead to either cell survival or death. Death receptors TNFR1 and Fas are type I membrane proteins with a conserved domain called death domain. Ligation of TNFR1 by TNF-α or Fas by Fas ligand (FasL) induces trimerization of the receptors resulting in the formation of a death-inducing signaling complex, which through activation of caspases 8 and 10, leads to apoptosis. The initiator caspases can either directly activate executioner caspase 3 or cleave Bid, a proapoptotic member of the Bcl-2 family. The truncated Bid translocates to the mitochondria stimulating cytochrome *c* release. Cytochrome *c*, together with Apaf-1, form the apoptosome, activating caspase 9, which in turn activates caspase 3. Then, caspase 3 and other effector caspases cleave substrates involved in cell disassembling [36].

TNF-α is synthesized in somatotropes and its receptors are expressed in the anterior pituitary gland [1, 37]. TNF-α release from anterior pituitary cells is higher at proestrus than at diestrus. Also, the inhibitory effect of TNF-α on anterior pituitary cell proliferation is predominant at proestrus and is estrogen-dependent [34]. TNF-α exerts an estrogen-dependent proapototic action on lactotropes and somatotropes [11, 13].

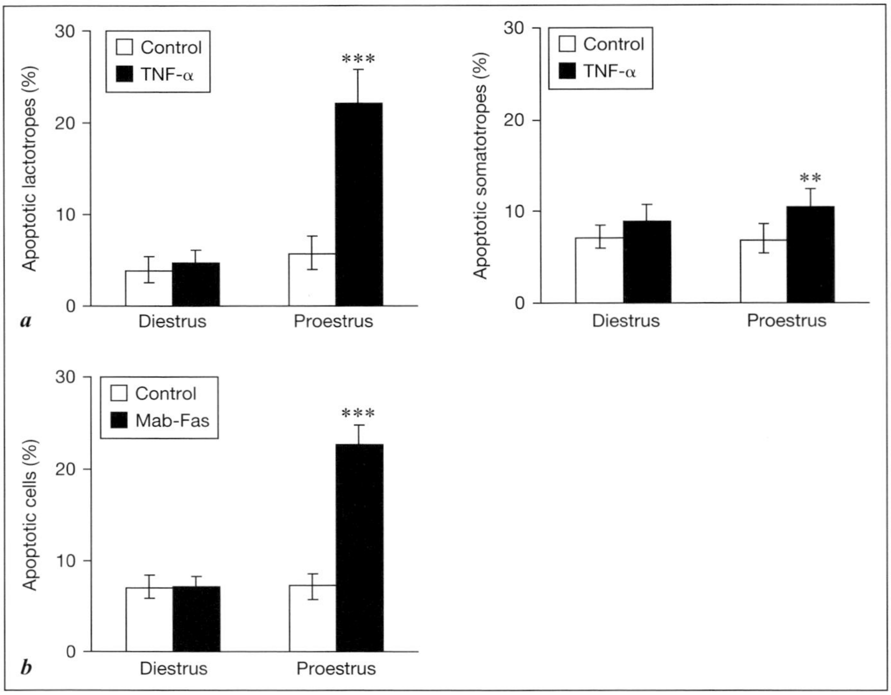

Fig. 1. Apoptosis in anterior pituitary cells from rats killed at selected stages of the estrous cycle. ***a*** Effect of TNF-α (50 ng/ml) on the percentage of apoptotic lactotropes and somatotropes. Each column represents the percentage ± CL of TUNEL-positive cells. ***b*** Effect of Fas activation on the percentage of apoptosis in anterior pituitary cells. Cells were incubated with an agonistic anti-Fas antibody (1 μg/ml, Mab-Fas). $^{**}p < 0.01$, $^{***}p < 0.001$ vs. respective controls (χ^2 test). Modified with permission from *Endocrinology* [11, 13, 38].

Fas and FasL are expressed in several pituitary cell types, mainly in lactotropes and somatotropes [38]. The expression of both Fas and FasL is higher in the anterior pituitary cells of rats at proestrus. Similarly to other estrogen-responsive tissues, the expression of the Fas/FasL system in anterior pituitary cells is upregulated by estradiol [38].

The apoptotic response of anterior pituitary cells to death receptor activation varies over the estrous cycle. Apoptosis induced by TNF-α is observed in lactotropes and somatotropes of rats at proestrus, whereas cells of rats at diestrus do not respond to this cytokine (fig. 1). Also, Fas activation induces apoptosis only in anterior pituitary cells of rats at proestrus but not at diestrus (fig. 1). In addition, the increase in the apoptotic index induced by lipopolysaccharide

(LPS) administration in the anterior pituitary gland is higher at proestrus than at other stages of the estrous cycle [39]. The highest apoptotic rate in the anterior pituitary gland at proestrus coincides with the highest expression of Bax, a proapoptotic member of the Bcl-2 family [14]. Therefore, anterior pituitary cells show cyclic changes in sensitivity to proapoptotic factors depending on the cyclic pattern of circulating gonadal hormones during the estrous cycle. Lactotropes show high sensitivity to proapoptotic stimuli at proestrus [11, 38], when circulating levels of estrogens are the highest [12].

Control of apoptosis in anterior pituitary cells by gonadal steroids could be exerted either directly or through regulation of the release of hypothalamic factors. It has been suggested that LHRH is involved in the inhibition of apoptosis at estrus [14]. Since dopamine receptor expression is highest at proestrus and dopamine is known to induce apoptosis of somatolactotrope GH3 cells, it could also be participating in the apoptosis of anterior pituitary cells at this stage of the estrous cycle [16, 40]. Nevertheless, gonadal steroids may modulate the processes of cell death in the anterior pituitary mainly by direct action. In fact, in primary cultures of anterior pituitary cells, TNFR and Fas activation induces apoptosis of lactotropes only when cells are incubated with 17β-estradiol (fig. 2). Interestingly, chronic estrogen treatment induces apoptosis in the anterior pituitary gland [39]. Also, in vitro studies showed that estradiol induces per se apoptosis of somatotropes (fig. 2). Although the mechanisms involved in the apoptotic action of estrogens in the anterior pituitary have yet to be elucidated, it was shown that chronic estrogenization increases the expression of tumor suppressor genes such as p53 [25]. Another mechanism that could be involved in the proapoptotic effect of estradiol is the modulation of the mitochondrial death regulation pathway. The expression of Bax is higher at proestrus, when estrogen levels are high [14]. Also, estradiol increases the expression of Bad, another proapoptotic member of this family, in the anterior pituitary gland [41].

Estradiol-induced apoptosis of somatotropes and the permissive action of estradiol on TNF-α-induced apoptosis on both lactotropes and somatotropes are antagonized by progesterone (fig. 2). Progesterone also inhibits some other actions of estrogens in the anterior pituitary, such as stimulation of cell proliferation and prolactin release [42]. Since the expression of progesterone receptors is confined to gonadotropes, these progesterone effects as well as their cytoprotective effect on lactotropes and somatotropes may be indirect and mediated by the paracrine action of growth factors and cytokines. Also, the antiapoptotic action of progesterone may involve changes in the distribution of estrogen receptors [42] and in the ratio of proapoptotic to antiapoptotic proteins of the Bcl-2 family [43].

Estradiol stimulates TNF-α release from anterior pituitary cells and enhances the stimulatory effect of LPS on TNF-α release suggesting that both

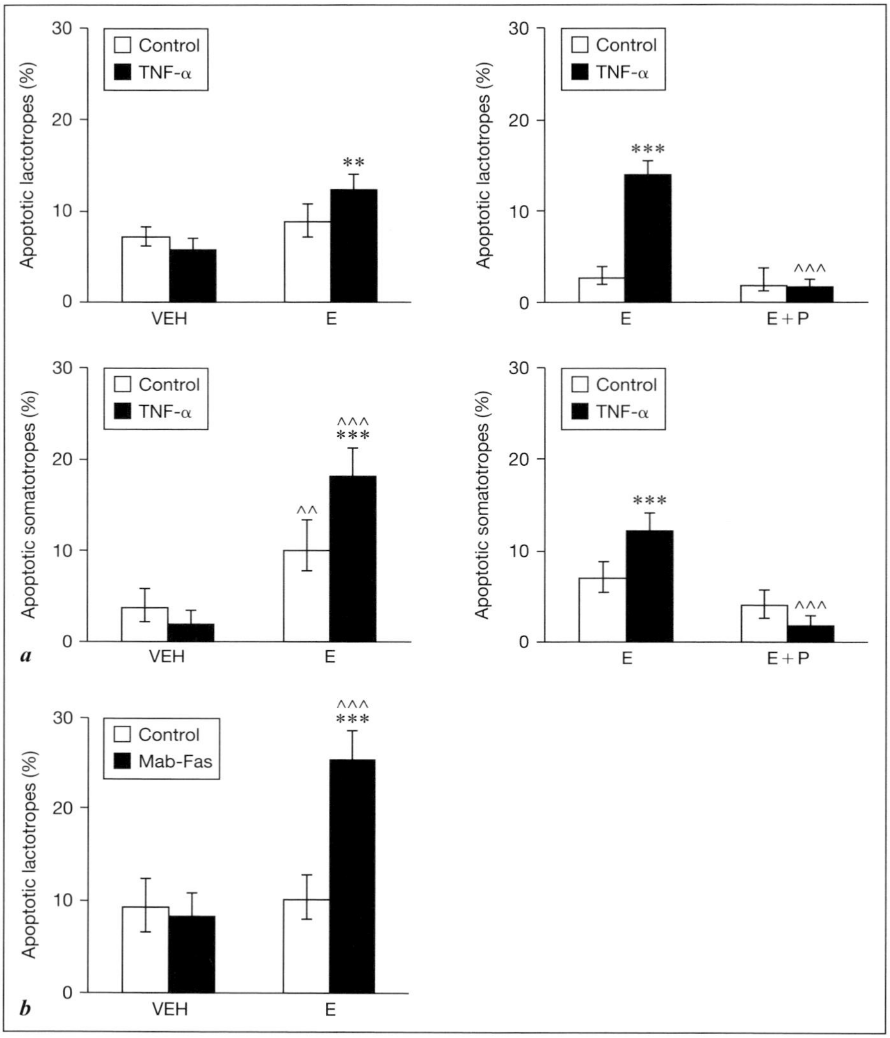

Fig. 2. Effect of gonadal steroids on TNF-α- and Fas-induced apoptosis of lactotropes and somatotropes of OVX rats. ***a*** Lactotropes and somatotropes of OVX rats cultured with vehicle (ethanol 1 μl/l, VEH), 17β-estradiol (10^{-9} *M*, E) or 17β-estradiol plus progesterone (10^{-6} *M*, E+P) were incubated in the presence of TNF-α (50 ng/ml). **$p < 0.01$, ***$p < 0.001$ vs. respective controls without TNF-α, ^^$p < 0.01$, ^^^$p < 0.001$ vs. respective controls without 17β-estradiol (χ^2 test). ***b*** Lactotropes from OVX rats cultured with vehicle or 17β-estradiol (10^{-9} *M*, E) were incubated in the presence of anti-Fas antibody (1 μg/ml, Mab-Fas). ***$p < 0.001$ vs. respective controls without Mab-Fas, ^^^$p < 0.001$ vs. respective controls without 17β-estradiol (χ^2 test). Modified with permission from *Endocrinology* [11, 13, 38].

basal and LPS induced TNF-α release from anterior pituitary cells is estrogen-dependent. Therefore, this gonadal steroid may be responsible for the higher levels of TNF-α release at proestrus [33]. Since estradiol induces apoptosis in the anterior pituitary gland and enables the proapoptotic action of LPS [39], estrogens may be involved in anterior pituitary cell renewal during the estrous cycle, sensitizing lactotropes, and probably other cell types, to proapoptotic factors.

Concluding Remarks

Estradiol stimulates cell proliferation and prevents apoptosis in many tissues but, in specific conditions, cells respond paradoxically to this gonadal steroid and undergo apoptosis [44]. Estrogens have classically been recognized as potent mitogens in the anterior pituitary gland. The majority of in vitro studies investigating direct actions of estradiol on cell proliferation, and also apoptosis, have used estrogen-responsive cell lines, so that the proliferative response in these cell lines may not represent the normal physiological response of anterior pituitary cells. Estrogens can trigger antiproliferative and apoptotic responses in lactotropes. Therefore, estradiol may participate in anterior pituitary cell renewal by sensitizing the cells not only to mitogenic stimuli but also to apoptotic signals. Cyclic expression of many hormones, growth factors and cytokines induced by estradiol during the estrous cycle may account for the rapid cyclical proliferative and apoptotic activity in anterior pituitary cells. The steroid milieu on the morning of proestrus could increase local production of auto-/paracrine proapoptotic factors and/or their receptors, which, in turn, would lead to apoptosis of anterior pituitary cells (fig. 3). Considering that the expression of TNF-α and the Fas/FasL system in anterior pituitary cells is higher at proestrus and stimulated by estradiol, it is possible that these mediators could participate in the process of anterior pituitary cell renewal during the estrous cycle, triggering apoptosis at proestrus. Also, the proapoptotic action of estradiol seems to be mediated by modifying phenotypic features that enhance the responsiveness of anterior pituitary cells to proapoptotic factors [44].

Progesterone may act as a survival factor in the anterior pituitary gland. On the morning of proestrus, when circulating levels of progesterone are still low, estrogen levels peak. This steroid profile may sensitize the anterior pituitary gland to proapoptotic signals at this stage of the estrous cycle. On the contrary, low circulating levels of estrogens and/or high progesterone levels may impair apoptosis of anterior pituitary cells at other stages of the estrous cycle (fig. 3).

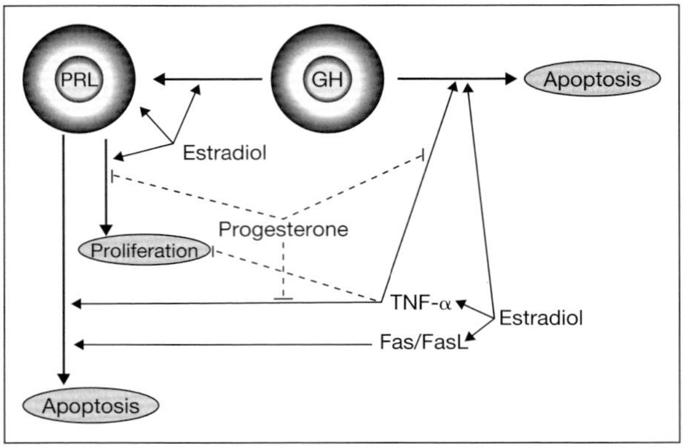

Fig. 3. Model for gonadal steroid regulation of proliferation and apoptosis of lactotropes (PRL) and somatotropes (GH). Estrogens stimulate lactotrope proliferation and the differentiation of somatotropes into lactotropes. They also enhance the expression of TNF-α and the Fas/FasL system and sensitize lactotropes to proapoptotic stimuli. Estradiol not only sensitizes somatotropes to the proapoptotic effect of TNF-α but also induces apoptosis in this cell type. Progesterone antagonizes both the mitogenic and proapoptotic action of estrogens. High circulating levels of estrogens and low levels of progesterone may trigger apoptosis of lactotropes at proestrus.

Even though the percentage of anterior pituitary cells renewed in each sexual cycle is low, dysregulation of this process during reproductive life in females may have consequences such as the generation of pituitary hyperplasia or tumors. Tumoral cells, such us GH3 and AtT20 undergo apoptotic cell death after overexpression of TNF-α and FasL [unpubl. results] suggesting that death signaling cascades triggered by TNFR1 and Fas are also present in tumoral pituitary cells. Considering that regulation of tissue homeostasis results from a balance between cell proliferation and death, and that cells with genetic alterations are physiologically eliminated by apoptosis, defects in cell death processes play an important role in tumor genesis and progression. Since apoptosis is a key event in the regulation of anterior pituitary cell renewal, death receptors could be considered as a target for the treatment of pituitary adenomas.

Acknowledgments

This work was funded by grants from the Agencia Nacional de Investigaciones Científicas y Tecnológicas, CONICET and Universidad de Buenos Aires, Argentina and by NIH grant 1RO3-TW006273–01A1.

References

1 Takahashi S: Intrapituitary regulatory system of proliferation of mammotrophs in the pituitary gland. Zoolog Sci 2004;21:601–611.

2 Denef C: Paracrine control of lactotrope proliferation and differentiation. Trends Endocrinol Metab 2003;14:188–195.

3 Yin P, Arita J: Differential regulation of prolactin release and lactotrope proliferation during pregnancy, lactation and the estrous cycle. Neuroendocrinology 2000;72:72–79.

4 Porter TE, Wiles CD, Frawley LS: Evidence for bidirectional interconversion of mammotropes and somatotropes: rapid reversion of acidophilic cell types to pregestational proportions after weaning. Endocrinology 1991;129:1215–1220.

5 Ahlbom E, Grandison L, Zhivotovsky B, Ceccatelli S: Termination of lactation induces apoptosis and alters the expression of the Bcl-2 family members in the rat anterior pituitary. Endocrinology 1998;139:2465–2471.

6 Nolan LA, Kavanagh E, Lightman SL, Levy A: Anterior pituitary cell proliferation control: basal cell turnover and the effects of adrenalectomy and dexamethasone treatment. J Neuroendocrinol 1998;10:207–215.

7 Oishi Y, Okuda M, Takahashi H, Fujii T, Morii S: Cellular proliferation in the anterior pituitary gland of normal adult rats: influences of sex, estrous cycle, and circadian change. Anat Rec 1993;235:111–120.

8 Hashi A, Mazawa S, Chen SY, Yamakawa K, Kato J, Arita J: Estradiol-induced diurnal changes in lactotroph proliferation and their hypothalamic regulation in ovariectomized rats. Endocrinology 1996;137:3246–3252.

9 Sasaki F, Iwama Y: Sex difference in prolactin and growth hormone cells in mouse adenohypophysis: stereological, morphometric, and immunohistochemical studies by light and electron microscopy. Endocrinology 1988;123:905–912.

10 Hashi A, Mazawa S, Kato J, Arita J: Pentobarbital anesthesia during the proestrous afternoon blocks lactotroph proliferation occurring on estrus in female rats. Endocrinology 1995;136:4665–4471.

11 Candolfi M, Zaldivar V, De Laurentiis A, Jaita G, Pisera D, Seilicovich A: TNF-alpha induces apoptosis of lactotropes from female rats. Endocrinology 2002;143:3611–3617.

12 Freeman ME: The neuroendocrine control of the ovarian cycle of the rat; in Knobil E, Neill JD (eds): The Physiology of Reproduction. New York, Raven Press, 1994, vol 2, pp 613–658.

13 Candolfi M, Jaita G, Zaldivar V, Zarate S, Ferrari L, Pisera D, Castro MG, Seilicovich A: Progesterone antagonizes the permissive action of estradiol on tumor necrosis factor-alpha-induced apoptosis of anterior pituitary cells. Endocrinology 2005;146:736–743.

14 Yin P, Arita J: Proestrous surge of gonadotropin-releasing hormone secretion inhibits apoptosis of anterior pituitary cells in cycling female rats. Neuroendocrinology 2002;76:272–282.

15 Markus GJ: Mitosis in the rat uterus during the estrous cycle, early pregnancy, and early pseudopregnancy. Biol Reprod 1974;10:447–452.

16 Sarkar DK, Chaturvedi K, Oomizu S, Bayadjieva NI, Chen CP: Dopamine, dopamine D2 receptor short isoform, transforming growth factor (TGF)-beta1, and TGF-beta type II receptor interact to inhibit the growth of pituitary lactotropes. Endocrinology 2005;146:4179–4188.

17 Shinkai T, Ooka H: Effect of angiotensin II on the proliferation of mammotrophs from the adult rat anterior pituitary in culture. Peptides 1995;16:25–29.

18 Arzt E, Pereda MP, Castro CP, Pagotto U, Renner U, Stalla GK: Pathophysiological role of the cytokine network in the anterior pituitary gland. Front Neuroendocrinol 1999;20:71–95.

19 Michels KM, Lee WH, Seltzer A, Saavedra JM, Bondy CA: Up-regulation of pituitary [^{125}I]insulin-like growth factor-I (IGF-I) binding and IGF binding protein-2 and IGF-I gene expression by estrogen. Endocrinology 1993;132:23–29.

20 Lara JI, Lorenzo MJ, Cacicedo L, Tolon RM, Balsa JA, Lopez-Fernandez J, Sanchez-Franco F: Induction of vasoactive intestinal peptide gene expression and prolactin secretion by insulin-like growth factor I in rat pituitary cells: evidence for an autoparacrine regulatory system. Endocrinology 1994;135:2526–2532.

21 Melmed S: Mechanisms for pituitary tumorigenesis: the plastic pituitary. J Clin Invest 2003;112: 1603–1618.

22 Chun TY, Wendell D, Gregg D, Gorski J: Estrogen-induced rat pituitary tumor is associated with loss of retinoblastoma susceptibility gene product. Mol Cell Endocrinol 1998;146:87–92.
23 Ying C, Gregg DW, Gorski J: Estrogen-induced changes in rRNA accumulation and RNA polymerase I activity in the rat pituitary: correlation with pituitary tumor susceptibility. Mol Cell Endocrinol 1996;118:207–213.
24 Mitchner NA, Garlick C, Ben-Jonathan N: Cellular distribution and gene regulation of estrogen receptors alpha and beta in the rat pituitary gland. Endocrinology 1998;139:3976–3983.
25 Ying C: Potential involvement of tumor suppressor gene expression in the formation of estrogen-inducible pituitary tumors in rats. Endocr J 2000;47:1–5.
26 Pandey J, Bannout A, Wendell DL: The Edpm5 locus prevents the 'angiogenic switch' in an estrogen-induced rat pituitary tumor. Carcinogenesis 2004;25:1829–1838.
27 Oomizu S, Chaturvedi K, Sarkar DK: Folliculostellate cells determine the susceptibility of lactotropes to estradiol's mitogenic action. Endocrinology 2004;145:1473–1480.
28 Lloyd RV, Landefeld TD, Maslar I, Frohman LA: Diethylstilbestrol inhibits tumor growth and prolactin production in rat pituitary tumors. Am J Pathol 1985;118:379–386.
29 Lamberts SW, Nagy I, Uitterlinden P, MacLeod RM: The effect of catecholestrogens on the growth of prolactin-secreting pituitary tumors and normal prolactin synthesis in the rat. Endocrinology 1982;110:1141–1146.
30 Lee EJ, Duan WR, Jakacka M, Gehm BD, Jameson JL: Dominant negative ER induces apoptosis in GH(4) pituitary lactotrope cells and inhibits tumor growth in nude mice. Endocrinology 2001;142:3756–3763.
31 Kawashima K, Yamakawa K, Takahashi W, Takizawa S, Yin P, Sugiyama N, Kanba S, Arita J: The estrogen-occupied estrogen receptor functions as a negative regulator to inhibit cell proliferation induced by insulin/IGF-1: a cell context-specific antimitogenic action of estradiol on rat lactotrophs in culture. Endocrinology 2002;143:2750–2758.
32 Yamakawa K, Arita J: Cross-talk between the estrogen receptor-, protein kinase A-, and mitogen-activated protein kinase-mediated signaling pathways in the regulation of lactotroph proliferation in primary culture. J Steroid Biochem Mol Biol 2004;88:123–130.
33 Theas MS, De Laurentis A, Lasaga M, Pisera D, Duvilanski BH, Seilcovich A: Effect of lipopolysaccharide on tumor necrosis factor and prolactin release from rat anterior pituitary cells. Endocrine 1998;8:241–245.
34 Theas S, Pisera D, Duvilanski B, De Laurentiis A, Pampillo M, Lasaga M, Seilicovich A: Estrogens modulate the inhibitory effect of tumor necrosis factor-alpha on anterior pituitary cell proliferation and prolactin release. Endocrine 2000;12:249–255.
35 Onofri C, Carbia Nagashima A, Schaaf L, Feirer M, Lohrer P, Stummer W, Berner S, Chervin A, Golldberg V, Stalla GK, Renner U, Arzt E: Estradiol stimulates vascular endothelial growth factor and interleukin-6 in human lactotroph and lactosomatotroph pituitary adenomas. Exp Clin Endocrinol Diabetes 2004;112:18–23.
36 Wajant H, Gerspach J, Pfizenmaier K: Tumor therapeutics by design: targeting and activation of death receptors. Cytokine Growth Factor Rev 2005;16:55–76.
37 Kobayashi H, Fukata J, Murakami N, Usui T, Ebisui O, Muro S, Hanaoka I, Inoue K, Imura H, Nakao K: Tumor necrosis factor receptor in the pituitary cells. Brain Res 1997;758:45–50.
38 Jaita G, Candolfi M, Zaldivar V, Zarate S, Ferrari L, Pisera D, Castro MG, Seilicovich A: Estrogens up-regulate the Fas/FasL apoptotic pathway in lactotropes. Endocrinology 2005;146:4737–4744.
39 Pisera D, Candolfi M, Navarra S, Ferraris J, Zaldivar V, Jaita G, Castro MG, Seilicovich A: Estrogens sensitize anterior pituitary gland to apoptosis. Am J Physiol Endocrinol Metab 2004;287:767–771.
40 Heiman ML, Ben-Jonathan N: Dopaminergic receptors in the rat anterior pituitary change during the estrous cycle. Endocrinology 1982;111:37–41.
41 Kulig E, Camper SA, Kuecker S, Jin L, Lloyd RV: Remodeling of hyperplastic pituitaries in hypothyroid us-subunit knockout mice after thyroxine and 1713-estradiol treatment: role of apoptosis. Endocr Pathol 1998;9:261–274.
42 Piroli GG, Grillo CA, Ferrini MG, Lux-Lantos V, De Nicola AF: Antagonism by progesterone of diethylstilbestrol-induced pituitary tumorigenesis in Fischer 344 rats: effects on sex steroid receptors and tyrosine hydroxylase mRNA. Neuroendocrinology 1996;63:530–539.

43 Pecci A, Scholz A, Pelster D, Beato M: Progestins prevent apoptosis in a rat endometrial cell line and increase the ratio of bcl-XL to bcl-XS. J Biol Chem 1997;272:11791–11798.
44 Song RX, Santen RJ: Apoptotic action of estrogen. Apoptosis 2003;8:55–60.

Adriana Seilicovich
Centro de Investigaciones en Reproducción, Facultad de Medicina
Universidad de Buenos Aires
Paraguay 2155, piso 10
Buenos Aires C1121ABG (Argentina)
Tel./Fax +54 11 48074052, E-Mail adyseili@fmed.uba.ar

Arzt E, Bronstein M, Guitelman M (eds): Pituitary Today: Molecular, Physiological and Clinical Aspects. Front Horm Res. Basel, Karger, 2006, vol 35, pp 22–31

Bone Morphogenetic Protein-4 Control of Pituitary Pathophysiology

Damiana Giacomini[a,], Marcelo Páez-Pereda[c], Marily Theodoropoulou[b], Juan Gerez[a], Alberto Carbia Nagashima[a,*], Alberto Chervin[d], Silvia Berner[d], Marta Labeur[b], Damian Refojo[a,*], Ulrich Renner[b], Günter K. Stalla[b], Eduardo Arzt[a,*]*

[a]Laboratorio de Fisiología y Biología Molecular, Departamento de Fisiología, Biología Molecular y Celular, FCEN, Universidad de Buenos Aires, Buenos Aires, Argentina; [b]Max-Planck Institute of Psychiatry, [c]Affectis Pharmaceuticals, Munich, Germany; [d]Department of Neurosurgery and Neuroendocrinology, Hospital Santa Lucía, Buenos Aires, Argentina

Abstract

Bone morphogenetic protein-4 (BMP-4), a member of the transforming growth factor-β (TGF-β) family, is overexpressed in different prolactinoma models and induces the development of these lineage adenomas. SMAD proteins activated by growth factors of the TGF-β and BMP family interact with estrogen receptors to stimulate the proliferation of prolactin and growth hormone-secreting cells. Furthermore, BMP-4 presents differential expression in normal and adenomatous corticotropes and inhibitory action on corticotropinoma cell proliferation. Moreover, BMP-4 mediates the antiproliferative action of retinoic acid in these cells. The present review highlights not only the crucial and opposite role of BMP-4 in the progression of pituitary adenomas but also that BMP-4 and retinoic acid interaction might serve as a potential new mechanism target for therapeutic approaches for Cushing disease.

The BMPs Family/TGF-β Superfamily

The bone morphogenetic proteins (BMPs) are members of the transforming growth factor-β (TGF-β) superfamily. TGF-β is a still growing superfamily of cytokines with widespread distribution and diverse biological functions. They fall into several subfamilies including the TGF-β1, 2 and 3, the BMPs, the

*Members of the IFYBINE, Argentine National Research Council (CONICET).

growth/differentiation factors (GDFs), the nodals, the activins and inhibins, the anti-müllerian hormone and members of the glial cell line-derived neurotrophic factor family [1, 2]. These factors regulate many different cellular events, including cell differentiation during embryonic development, cell adhesion, migration, proliferation, death and transformation throughout the life span of the organism [1, 2]. For this reason the TGF-β superfamily signaling pathway is under intense investigation, including the pituitary where it exert an important regulatory role.

The signal transduction pathway of the superfamily begins with the ligand binding to its specific receptor. There are two different transmembrane protein serine/threonine kinases known as receptor types I and II, and upon ligand binding to receptor type II these two receptors are brought together. In this ligand-induced complex the receptor II phosphorylates receptor type I kinase, resulting in its activation. Thus, this is the initial step during intracellular signaling. Receptor type I activation allows the signal propagation through recognition and phosphorylation of the Smad subgroup known as receptor-activated Smads (R-Smads). This subgroup includes Smad2 and Smad3 which are specifically recognized by TGF-β and activin receptors and Smad1, Smad5 and Smad8 activated by BMP receptors. Upon phosphorylation the R-Smads undergo homotrimerization and formation of heteromeric complexes with the Co-Smad, Smad4 that is the only Smad protein common to all R-Smad [3]. The activated Smad complexes then translocated into the nucleus and in association with other nuclear cofactors bind to target sequences to regulate gene transcription. Another subgroup of Smad proteins are the I-Smads, Smad6 and Smad7, which compete with R-Smads for receptor or Co-Smad interaction or by inducing receptor degradation, thus the I-Smads negatively regulate TGF-β signaling [3].

Other mechanisms to inhibit or downregulate TGF-β signaling also exist. Receptor activation is regulated by a group of soluble proteins, which have the ability to bind TGF-β family factors and prevent the binding to their receptors thus controlling the extracellular phase of the TGF-β pathway. This group includes the latency-associated protein (LAP), the small proteoglycan decorin and the circulating protein α_2-macroglobulin, which all bind free TGF-β; follistatin which binds to activins and BMPs; and noggin, chordin and the related factors caronte, cerberus and gremlin that specifically bind BMP members [4]. Receptor activation is also inhibit by the decoy receptor BAMBI, this protein competes with the type I receptor for ligand-induced receptor dimerization. The intracellular protein FKBP12 also binds to the unphosphorylated type I receptor preventing its activation. Furthermore, dephosphorylation in addition to ubiquitination and proteasome-mediated degradation of the activated R-Smads and TGF-β receptors (mediated by Smurf1 and Smurf2) are two known mechanisms to shut down TGF-β signaling. As mentioned above, the inhibitory Smads, Smad6 and Smad7, play an important role in this signaling inactivation process. Smad7 directly binds to the activated type I TGF-β receptor upon TGF-β and BMP stimulation and inhibits phosphorylation of R-Smads. On the other hand, Smad6 specifically inhibit BMP signaling by competing with the activated Smad1 for Smad4 binding, it serves as a Smad4 decoy [reviewed in 3].

BMP family members have been originally studied by their ability to induce ectopic cartilage and bone formation when injected in experimental animals models [5]. Today, it is well known that that they not only promote differentiation of bone lineage cells [6, 7] but also regulate cellular division, apoptosis, cellular determination, differentiation and morphogenesis [reviewed in 8].

Because of its participation in several embryonic events, BMP-4 is one of the most studied members of the BMP family. BMP-4 is known to be involved in mesoderm development and cellular commitment during and after the gastrulation process [9, 10]. This important role during development has been shown by studying BMP-4 spatial and temporal expression in mouse embryos, and it has been observed that BMP-4 expression is indeed induced in future mesodermic structures [11]. Moreover, the BMP-4 knock-out mouse supported this result as this mutation is lethal before or during the gastrulation process and there is a lack of embryonic mesoderm differentiation in these mice [10].

At the organogenesis level BMP-4 (among with other BMPs) is involved in epidermis differentiation and inhibits the induction of neuronal ectodermic tissue. Experiments in *Xenopus* embryos have demonstrated that during neuronal development BMP-4 expression is lost in the presumptive neuronal tissue and is only detected in the epidermis [12, 13]. Moreover, a dominant negative form for BMP-4 or for the BMP-4 receptor promotes the differentiation of neuronal tissue in the ectoderm during *Xenopus* development [12, 14], supporting the negative role of BMP-4 during neuronal organogenesis. Noggin, cordina and folistatina are known factors that play an important role in neuronal differentiation by the inhibitory action of BMP-4 [15, 16]. BMP-4 is also involved during other tissue organogenesis, e.g. BMP-4 promotes tissue development in lungs, liver, kidney, the urinary system and teeth [8, 17–19].

Expression of TGF-β Superfamily Factors/Receptors during Pituitary Development

Several members of the TGF-β family are expressed in the pituitary gland at different stages of differentiation where they have crucial roles. During pituitary organogenesis BMP-2 and BMP-4 have been shown to play a role in the initial steps of the development of the anterior pituitary [20]. For instance in a first stage, BMP-4 is required for the proliferation of the Rathke pouch placode, which gives rise to prolactin-secreting cells, among others. The overexpression of noggin or a dominant negative BMP receptor (BMPR1A) in the anterior pituitary leads to an arrest in the development of prolactin-secreting cells [20]. During the second pituitary organogenesis stage, an inhibition of BMP-2 ventrodorsal gradient by an FGF8 dorsoventral gradient leads to corticotrope differentiation [21–24]. Overexpression of FGF8 in the developmental pituitary results in the absence of Pit-1 lineage cells and enhanced corticotrope differentiation, suggesting that it is necessary to inhibit BMP signaling for normal corticotrope development [24, 25]. Thus, regulation of BMP signaling is an important event for the proper organogenesis of pituitary lineage cells.

BMP-4 Expression and Action in Adult Pituitary Cells

As a first step in order to elucidate the mechanisms leading to pituitary pathogenesis we compared, by mRNA differential display PCR, the gene expression pattern of normal anterior pituitaries versus prolactinomas [26]. We observed that the BMP extracellular inhibitor noggin was downregulated in spontaneously developed prolactinomas taken from dopamine D2-receptor-deficient (D2R−/−) female mice, as compared to the normal pituitary of wild-type mice. On the other hand, BMP-4 was overexpressed in prolactinomas not only from the D2R−/− mice but also in other prolactinoma models, including estradiol-induced rat prolactinomas and human prolactinomas, as compared to normal tissue and other pituitary adenoma types. Also, we found that BMP-4 stimulated cell proliferation in vitro and promoted expression of the oncogene c-Myc in human prolactinomas, while it had no action in other human pituitary tumors. Given that Smads proteins mediate BMP-4 signaling and to further study the role of BMP-4 in tumor formation in nude mice, GH3 clones stably transfected with a Smad-4 dominant-negative (GH3-Smad-4dn) protein were produced and injected into nude mice. GH3-Smad-4dn cells formed small slow-growing tumors that did not express c-Myc as compared to control cells. Tumor growth and c-Myc expression were recovered when the Smad-4dn expression was lost, proving direct evidence that BMP-4/Smad-4 are involved in tumor development in vivo. These results also showed that BMP-4-positive growth signaling in the control of PRL-secreting cells overrides the negative effect of TGF-β, since GH3 Smad-4dn cells (which do not respond to either BMP-4 or TGF-β) have a blunted growth in nude mice. Furthermore, cell proliferation was upregulated by an overlapping intracellular signaling mechanism between BMP-4 and estrogens, and that their action was partially inhibited by blocking either pathway with the reciprocal antagonist, showing that BMP-4 effects are partially mediated by the estrogen receptor and that conversely, estradiol effects are partially mediated by Smad-4. Futhermore, by co-immunoprecipitation studies it was demonstrated that Smad-4 physically interacts with the ER [26]. The fact that the reciprocal antagonists only have a partial effect suggests that their cross-talk is not the exclusive pathway and that both BMP-4 and estradiol stimulate cell proliferation by other parallel mechanisms. But the fact that antiestrogens partially inhibit the effects of BMP-4 in prolactinoma cells suggests that the cross-talk between BMP-4 and estrogens could be a new target for specific antiestrogenic drugs. Moreover, the physical interaction between Smad-4 and ER may be present not only in prolactinoma cells but also in other cells such as breast tumors and bone, in which both estrogens and the TGF-β family play important roles [27, 28]. A physical interaction between ER and TGF-β-induced Smad-3 has also been recently shown in a reconstituted TGF-β and ER signaling system in the 293T cell line [29]. On the basis of these results, we propose that prolactinomas, in which BMP-4 stimulates cell proliferation, may be an interesting model to investigate the mechanisms that regulates the escape from the inhibition of cell proliferation by TGF-β family members in tumor cells.

Taking the crucial role of BMP-4 during pituitary organogenesis and prolactinoma tumorigenesis into consideration, we recently decided to study its potential action in the growth of corticotropinomas. Using AtT-20 corticotropinoma cells as a model, we demonstrated that BMP-4 inhibited both endogenous ACTH production and forskolin-induced ACTH stimulation in a dose-dependent manner and this inhibitory effect was blocked by noggin [30]. These results are in agreement with recently published data showing that BMP-4 has an inhibitory effect on the transcriptional activity of the POMC promoter [31]. To further establish the actions of BMP-4 on the physiology of corticotrope cells, we studied cell proliferation. BMP-4 inhibited not only AtT-20 cell proliferation but also c-myc promoter transcriptional activity. Based not only on these results and our previous ones but also on the fact that different studies that place the TGF-β family as important regulating factors in pituitary tumor development exist [32, 33], we studied a possible role for BMP-4 in the tumorigenicity of corticotrope cells in vivo. Considering the BMP-4 inhibitory action in vitro, we hypothesized that blocking BMP-4 inhibitory growth control would result in enhanced tumor growth. Thus, in order to observe differences in tumor growth, nude mice were injected with a suboptimal concentration of AtT-20 cells stably transfected with an empty vector, Smad-4dn or with noggin. At low cell concentration doses AtT-20 cells present no tumorigenic capacity when injected into nude mice. In contrast, AtT-20 Smad-4dn and AtT-20 noggin clones, in which the BMP-4 signaling pathway is blocked, developed visible tumors starting at 1 week after injection as well as symptoms of Cushing disease.

Therefore, inhibition of the BMP-4 signaling results in an increase of the AtT-20 cells tumorigenic potential demonstrating an inhibitory role of BMP-4 in the development of corticotrope tumors in vivo [30].

In agreement with BMP-4 inhibitory function during the development of corticotropinomas, BMP-4 expression was detected in 40% of endocrine cells in the normal adenohypophysis and most of the BMP-4 immunopositive cells were somatotropes and corticotropes but not lactotropes. On the other hand, there was a reduction of BMP-4 expression in corticotropinomas. Thus, BMP-4, a corticotrope growth inhibitor was expressed in the normal pituitary but was reduce in corticotrope tumors. This imbalance may allow growth promotion in corticotrope tumors.

Role of BMP-4 in Mediating Retinoic Acid Effects on Corticotropes

Recently, we have demonstrated that retinoic acid treatment reduced proopiomelanocortin (POMC) gene transcription and ACTH production by inhibiting the AP-1 and Nur77/Nurr1 transcriptional activities in pituitary ACTH-secreting tumor cells [34]. Also, in vivo experiments have shown that retinoic acid administration completely blocks the growth of corticotrope tumors and reverses the endocrine alterations and symptoms of Cushing disease [34]. In AtT-20 corticotrope cells retinoic acid treatment induces not only BMP-4 transcriptional activity but also BMP-4 expression and expression of COUP-TFI, an inhibitor of retinoic acid response pathways [35–38], blocks the stimulatory effect of retinoic acid. In cell proliferation assays both BMP-4 and retinoic acid inhibit AtT-20 cell proliferation, but there was no further inhibition in the presence of both factors even at low doses suggesting that they may share an overlapping (saturated at low doses) mechanism. As mentioned above, a cross-talk mechanism between BMP-4 and estrogens that regulates cell proliferation and a physical interaction between steroid receptors and Smads proteins actually exists [26, 29, 39] so an interaction between Smad proteins and the retinoic acid receptor could not be excluded. Also several studies have also established a different interaction between retinoic acid and BMPs [40–42] and considering that retinoic acid induces BMP-4 transcriptional activity and protein levels, the inhibitory effect of retinoic acid may be through BMP-4 induction. In AtT-20 noggin and AtT-20 Smad4-dn, the inhibitory effect of retinoic acid is abolished so we can conclude that retinoic acid, besides the possibility of a further crosstalk with BMP-4 signaling pathway, is acting through the induction of BMP-4 expression, which in turn has a central role in preventing corticotrope cell proliferation and tumoral cell growth [30].

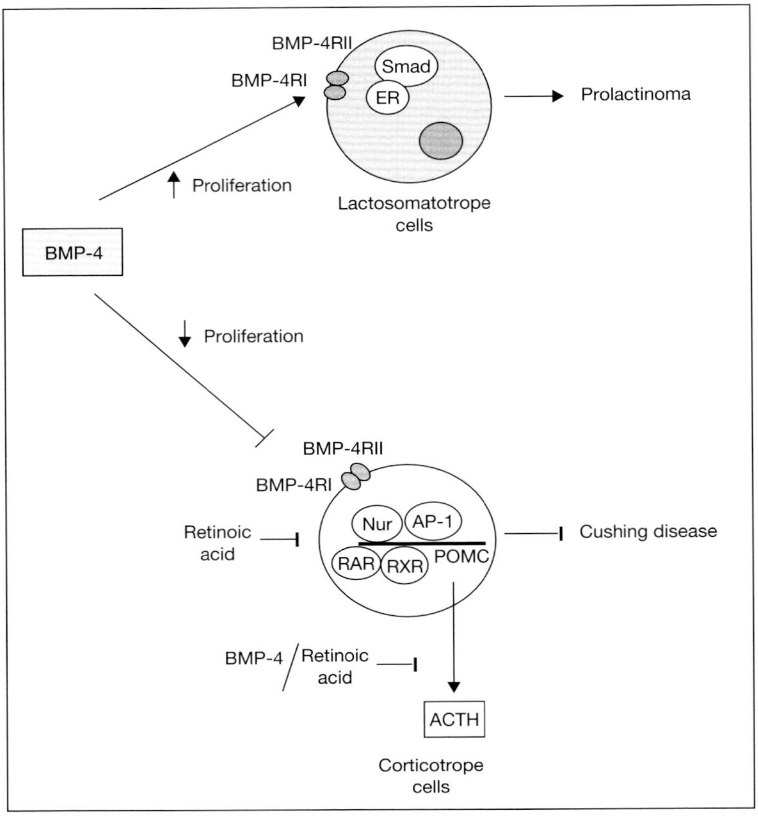

Fig. 1. BMP-4 promotes pituitary prolactinoma pathogenesis through a Smad/estrogen receptor crosstalk but inhibits ACTH production, corticotrope pathogenesis and Cushing disease, highlighting the important role of BMP-4 in the development of pituitary tumors in vivo. Retinoic acid also inhibits corticotrope cell physiology and induces BMP-4 transcription and expression so that its antiproliferative effect is mediated in part through BMP-4 induction.

Conclusions

The pathogenic mechanisms leading to prolactinomas and development of Cushing disease still remains an open question. We have demonstrated that BMP-4, which is crucial during pituitary organogenesis, has an opposite role in the different pituitary lineage adenomas (fig. 1). BMP-4 promotes the pathogenesis of pituitary prolactinomas through a Smad/estrogen receptor crosstalk but inhibits the pathogenesis of corticotrope cell proliferation and Cushing disease.

At this stage, it remains to be established which is the trigger for BMP-4 action (enhancement of proliferation in prolactinomas and loss of inhibition control in Cushing disease corticotrope proliferation) in adult pituitary tumor cells. Thus, BMP-4 acts as a regulator not only of pituitary cells undergoing differentiation of organogenesis, but also of transformed, not normal, adult pituitary cells, and it has a pituitary linage cell specificity during the tumorigenic process.

Acknowledgements

This work was supported by grants from the University of Buenos Aires, the Argentine National Research Council (CONICET) and Agencia Nacional de Promoción Científica y Tecnológica, Argentina.

References

1 Massague J: TGF-beta signal transduction. Annu Rev Biochem 1998;67:753–791.
2 Massague J: How cells read TGF-beta signals. Nat Rev Mol Cell Biol 2000;1:169–178.
3 Shi Y, Massague J: Mechanisms of TGF-beta signaling from cell membrane to the nucleus. Cell 2003;113:685–700.
4 Massague J, Chen YG: Controlling TGF-beta signaling. Genes Dev 2000;14:627–644.
5 Wozney JM, Rosen V, Celeste AJ, Mitsock LM, Whitters MJ, Kriz RW, Hewick RM, Wang EA: Novel regulators of bone formation: molecular clones and activities. Science 1988;242: 1528–1534.
6 Yamaguchi A, Katagiri T, Ikeda T, Wozney JM, Rosen V, Wang EA, Kahn AJ, Suda T, Yoshiki S: Recombinant human bone morphogenetic protein-2 stimulates osteoblastic maturation and inhibits myogenic differentiation in vitro. J Cell Biol 1991;113:681–687.
7 Duprez DM, Coltey M, Amthor H, Brickell PM, Tickle C: Bone morphogenetic protein-2 (BMP-2) inhibits muscle development and promotes cartilage formation in chick limb bud cultures. Dev Biol 1996;174:448–452.
8 Bellusci S, Henderson R, Winnier G, Oikawa T, Hogan BL: Evidence from normal expression and targeted misexpression that bone morphogenetic protein (Bmp-4) plays a role in mouse embryonic lung morphogenesis. Development 1996;122:1693–1702.
9 Hogan BL, Blessing M, Winnier GE, Suzuki N, Jones CM: Growth factors in development: the role of TGF-beta related polypeptide signalling molecules in embryogenesis. Dev Suppl 1994:53–60.
10 Winnier G, Blessing M, Labosky PA, Hogan BL: Bone morphogenetic protein-4 is required for mesoderm formation and patterning in the mouse. Genes Dev 1995;9:2105–2116.
11 Jones CM, Lyons KM, Hogan BLM: Involvement of bone morphogenetic protein (Bmp-4) and Vgr-1 in morphogenesis and neurogenesis in mouse. Development 1991;111:531–542.
12 Wilson PA, Hemmati-Brivanlou A: Induction of epidermis and inhibition of neural fate by Bmp-4. Nature 1995;376:331–333.
13 Harland RM: Neural induction in *Xenopus*. Curr Opin Genet Dev 1994;4:543–549.
14 Hawley SH, Wunnenberg-Stapleton K, Hashimoto C, Laurent MN, Watabe T, Blumberg BW, Cho KWY: Disruption of BMP signals in embryonic Xenopus ectoderm leads to direct neural induction. Genes Dev 1995;9:2923–2935.
15 Lamb TM, Knecht AK, Smith WC, Stachel SE, Economides AN, Stahl N, Yancopolous GD, Harland RM: Neural induction by the secreted polypeptide noggin. Science 1993;262:713–718.
16 Sasai Y, Lu B, Steinbeisser H, De Robertis EM: Regulation of neural induction by the Chd and Bmp-4 antagonistic patterning signals in *Xenopus*. Nature 1995;377:757.

17 Rossi JM, Dunn NR, Hogan BL, Zaret KS: Distinct mesodermal signals, including BMPs from septum transversum mesenchyme, are required in combination for hepatogenesis from the endoderm. Genes Dev 2001;15:1998–2009.
18 Dudley AT, Lyons KM, Robertson EJ: A requirement for bone morphogenetic protein-7 during development of the mammalian kidney and eye. Genes Dev 1995;9:2795–2807.
19 Roberts DJ, Johnson RL, Burke AC, Nelson CE, Morgan BA, Tabin C: Sonic hedgehog is an endodermal signal inducing Bmp-4 and Hox genes during induction and regionalization of the chick hindgut. Development 1995;121:3163–3174.
20 Scully KM, Rosenfeld MG: Pituitary development: regulatory codes in mammalian organogenesis. Science 2002;295:2231–2235.
21 Dasen JS, Rosenfeld MG: Signaling mechanisms in pituitary morphogenesis and cell fate determination. Curr Opin Cell Biol 1999;11:669–677.
22 Dasen JS, Rosenfeld MG: Signaling and transcriptional mechanisms in pituitary development. Annu Rev Neurosci 2001;24:327–355.
23 Treier M, Gleiberman AS, O'Connell SM, Szeto DP, McMahon JA, McMahon AP, Rosenfeld MG: Multistep signaling requirements for pituitary organogenesis in vivo. Genes Dev 1998;12: 1691–1704.
24 Kioussi C, Carriere C, Rosenfeld MG: A model for the development of the hypothalamic-pituitary axis: transcribing the hypophysis. Mech Dev 1999;81:23–35.
25 Kioussi C, O'Connell S, St-Onge L, Treier M, Gleiberman AS, Gruss P, Rosenfeld MG: Pax6 is essential for establishing ventral-dorsal cell boundaries in pituitary gland development. Proc Natl Acad Sci USA 1999;96:14378–14382.
26 Paez-Pereda M, Giacomini D, Refojo D, Nagashima AC, Hopfner U, Grubler Y, Chervin A, Goldberg V, Goya R, Hentges ST, Low MJ, Holsboer F, Stalla GK, Arzt E: Involvement of bone morphogenetic protein 4 (BMP-4) in pituitary prolactinoma pathogenesis through a Smad/estrogen receptor crosstalk. Proc Natl Acad Sci USA 2003;100:1034–1039.
27 Karsenty G: The genetic transformation of bone biology. Genes Dev 1999;13:3037–3051.
28 Massague J, Blain SW, Lo RS: TGFbeta signaling in growth control, cancer, and heritable disorders. Cell 2000;103:295–309.
29 Matsuda T, Yamamoto T, Muraguchi A, Saatcioglu F: Cross-talk between transforming growth factor-beta and estrogen receptor signaling through Smad3. J Biol Chem 2001;276:42908–42914.
30 Giacomini D, Paez-Pereda M, Theodoropoulou M, Labeur M, Refojo D, Gerez J, Chervin A, Berner S, Losa M, Buchfelder M, Renner U, Stalla GK, Arzt E: Bone morphogenetic protein-4 inhibits corticotroph tumor cells: involvement in the retinoic acid inhibitory action. Endocrinology 2006;147: 247–256.
31 Nudi M, Ouimette JF, Drouin J: Bone morphogenic protein (Smad)-mediated repression of proopiomelanocortin transcription by interference with Pitx/Tpit activity. Mol Endocrinol 2005;19: 1329–1342.
32 Hentges S, Sarkar DK: Transforming growth factor-beta regulation of estradiol-induced prolactinomas. Front Neuroendocrinol 2001;22:340–363.
33 Sheng HZ, Moriyama K, Yamashita T, Li H, Potter SS, Mahon KA, Westphal H: Multistep control of pituitary organogenesis. Science 1997;278:1809–1812.
34 Páez-Pereda M, Kovalovsky D, Hopfner U, Theodoropoulou M, Pagotto U, Uhl E, Losa M, Stalla J, Grubler Y, Missale C, Arzt E, Stalla G: Retinoic acid prevents experimental Cushing syndrome. J Clin Invest 2001;108:1123–1131.
35 Tran P, Zhang XK, Salbert G, Hermann T, Lehmann JM, Pfahl M: COUP orphan receptors are negative regulators of retinoic acid response pathways. Mol Cell Biol 1992;12:4666–4676.
36 Kliewer SA, Umesono K, Heyman RA, Mangelsdorf DJ, Dyck JA, Evans RM: Retinoid × receptor-COUP-TF interactions modulate retinoic acid signaling. Proc Natl Acad Sci USA 1992;89: 1448–1452.
37 Wu Q, Li Y, Liu R, Agadir A, Lee MO, Liu Y, Zhang X: Modulation of retinoic acid sensitivity in lung cancer cells through dynamic balance of orphan receptors nur77 and COUP-TF and their heterodimerization. EMBO J 1997;16:1656–1669.
38 Beland M, Lohnes D: Chicken ovalbumin upstream promoter-transcription factor members repress retinoic acid-induced Cdx1 expression. J Biol Chem 2005;280:13858–13862.

39 Kang HY, Huang KE, Chang SY, Ma WL, Lin WJ, Chang C: Differential modulation of androgen receptor-mediated transactivation by Smad3 and tumor suppressor Smad4. J Biol Chem 2002;277:43749–43756.
40 Skillington J, Choy L, Derynck R: Bone morphogenetic protein and retinoic acid signaling cooperate to induce osteoblast differentiation of preadipocytes. J Cell Biol 2002;159:135–146.
41 Sumantran VN, Brederlau A, Funa K: BMP-6 and retinoic acid synergistically differentiate the IMR-32 human neuroblastoma cells. Anticancer Res 2003;23:1297–1303.
42 Li X, Schwarz EM, Zuscik MJ, Rosier RN, Ionescu AM, Puzas JE, Drissi H, Sheu TJ, O'Keefe RJ: Retinoic acid stimulates chondrocyte differentiation and enhances bone morphogenetic protein effects through induction of Smad1 and Smad5. Endocrinology 2003;144:2514–2523.

Dr. Eduardo Arzt
Laboratorio Fisiología y Biología Molecular
Depaptamento de Fisiología y Biología Molecular y Celular
FCEN, Universidad de Buenos Aires
Ciudad Universitaria, Pabellon II
1428 Buenos Aires (Argentina)
Tel. +54 11 45763368, Fax +54 11 45763321, E-Mail earzt@fbmc.fcen.uba.ar

Arzt E, Bronstein M, Guitelman M (eds): Pituitary Today: Molecular, Physiological and Clinical Aspects. Front Horm Res. Basel, Karger, 2006, vol 35, pp 32–49

Genesis of Prolactinomas: Studies Using Estrogen-Treated Animals

Dipak K. Sarkar

Endocrine Program, Biomedical Division of the Center of Alcohol Studies and Department of Animal Sciences, Rutgers, State University of New Jersey, New Brunswick, N.J., USA

Abstract

Prolactin-secreting adenomas (prolactinomas) are the most prevalent form of pituitary tumors in humans. Our knowledge of the formation of these tumors is limited. Experimental work in animal has uncovered that estradiol exposure leads to prolactinoma formation via orchestrated events involving dopamine D2 receptors, transforming growth factor-β (TGF-β) isoforms and their receptors, as well as factors secondary to TGF-β action. Additionally, these studies determined that TGF-β and b-FGF interact to facilitate the communication between lactotropes and folliculo-stellate cells that is necessary for the mitogenic action of estradiol. The downstream signaling that governs lactotropic cell proliferation involves activation of the MAP kinase p44/42-dependent pathway.

Pituitary tumors are primarily adenomas; they account for approximately 10–15% of intracranial tumors. They cause significant morbidity due to local invasion, hypopituitarism or hormone hypersecretion [1, 2]. Pituitary tumors are classified as either functioning and secreting excessive amounts of active hormones, or as endocrinologically inactive and secreting no hormones or inactive hormones. Pituitary tumors are also classified by virtue of their size into the arbitrary division of those less than 1 cm as microadenomas and those more than 1 cm as macroadenomas. Pituitary tumors secreting excess prolactin are characterized as prolactinomas. Prolactinomas are the most frequently occurring neoplasm in the human pituitary [3, 4]. In the general population, 1:2,800 men and 1:1,050 women are considered to have prolactinomas [5]. In human subjects, prolactinomas occur as both macroadenomas and microadenomas. In addition, mixed growth hormone and prolactin-secreting adenomas are documented to exist in a substantial number of acromegaly patients.

Hyperprolactinoma is a condition in which plasma prolactin (PRL) levels are elevated above normal levels. Hyperprolactinemia, with elevation of serum prolactin of more than 200 ng/ml, is characteristically associated with prolactinomas [6]. Hyperprolactinemia causes reproductive dysfunction such as amenorrhea, galactorrhea, and infertility in women [7]. Amenorrhea and galactorrhea may occur alone or together [8]. Up to 25% of patients with secondary amenorrhea have been diagnosed with hyperprolactinemia. Many of these patients showed micro-prolactin adenomas or macro-prolactin adenomas in the pituitary. Although treatments that alter central dopaminergic neuronal functions cause an elevated serum PRL level, in most cases hyperprolactinemia is due to a pituitary tumor. In women, prolactinomas are mainly microadenomas. These microadenomas are rarely associated with hypopituitarism or central nervous system dysfunction. Idiopathic hyperprolactinemia, without demonstrable pituitary or hypothalamic disease, has also been identified. In men, prolactinomas are mainly macroadenomas [9]. These patients often exhibit gynecomastia, impotence, decreased libido, and reduced reproductive hormone levels [10].

The genesis of prolactinoma in lactotropes of the pituitary gland or other pituitary tumors remains for the most part a subject of speculation and debate because cells involved in the formation of adenoma may be derived from monoclonal or polyclonal expansion and may have phenotypes that change during tumor development and therefore are very hard to discern. However, information of genetic modifications that increase the risk factors for lactotropic tumors has begun to decipher from animal studies in which prolactinoma is induced by estrogen.

Animal Model for Studying the Genesis of Prolactinomas

Aging Rats and Mice

Prolactinomas are not only the most frequently occurring neoplasm in the human pituitary, but these tumors also commonly occur in laboratory animals. A 10–86% incidence of spontaneous pituitary adenomas was reported in male and old female rats of different strains, especially during aging [reviewed in 11]. The incidences of pituitary adenomas in aging Wistar male and female rats, Charles River COBS male rats, and Long-Evans female rats are 56%, 10–51% and 56%, respectively. Most of these adenomas in laboratory rats are prolactinomas. Evidence for aging-induced prolactinomas in humans is currently lacking.

Estrogen-Treated Rats

Pituitary tumors in experimental animals can be induced by estrogen. Most of these estrogen-induced pituitary tumors are PRL- or GH-secreting tumors [12]. In both sexes of rats, long-term elevation of serum estrogen causes hyperplasia and/or adenomas [13, 14]. In Fischer-344 female rats, maintenance of constant elevated systemic estradiol level within the range of 100–300 pg/ml by subcutaneously implanting a 17β-estradiol-containing capsule induces prolactinomas very rapidly; within 2 weeks of estrogen implantation, significant hyperplasic response of lactotropes is evident [11, 13, 15, 17]. Estradiol administration also induces prolactinomas both in ovary intact or ovariectomized ACI rats [16]. Synthetic estrogen diethylstilbestrol also causes pituitary hyperplasia and neoplasm in Fischer 344 rats [17].

Estrogen also appears to increase the risk factors for prolactinomas in humans. There are reports of the development of prolactinemia in a male to female transsexual receiving massive doses of estrogen [19]. There is evidence of growth of microprolactinoma to a macroprolactinoma during estrogen therapy [20]. Women taking oral estrogen contraceptive showed higher PRL levels [21]. Women using oral contraceptive due to menstrual irregularities showed a 7- to 8-fold higher incidence of prolactinomas [22]. Data suggest that some women are more sensitive to the lactogenic effects of exogenous estrogen and, therefore, may be at greater risk for developing prolactinomas [23]. During pregnancy, estrogen levels elevate and the number of prolactin-secreting cells and serum PRL content increases. These elevated levels of estrogen are associated with symptomatic pituitary tumor enlargement in up to 30% of women with macroadenomas and may result in persistent hyperprolactinemia and postpartum amenorrhea or galactorrhea [24]. However, the risk for development of significant clinical symptoms related to tumor expansion is less than 2% in pregnant women with microadenomas [9]. There appears to be a close association between the use of oral estrogenic contraceptives and the onset of amenorrhea, often accompanied by galactorrhea [25–27]. Certain subsets of humans and animals are more sensitive to estrogen's mitogenic properties than the general population [28, 29]. Hence, estrogen can be considered a risk factor for the development of prolactinomas in some laboratory animals and in a certain population of humans. At present, there is no answer as to why some humans and animals are more responsive to estrogen's actions. However, it is interesting to note that there are slight structural differences in the pituitaries of estrogen-sensitive Fischer-344 and estrogen-insensitive Sprague-Dawley rats, including a difference in the number of supporting folliculo-stellate cells [30]. The differences in folliculo-stellate cell numbers between these two strains could be significant in tumorigenesis because these cells are a major source of bFGF, which regulates lactotropic cell proliferation [31].

Alcohol-Treated Rats

Several reports show evidence for the existence of high levels of PRL in chronic alcoholic men and women. In a study conducted by European scientists, include organochlorine pollutants such as PCBs and DDT, which are known to persist in the environment. Experimental studies have confirmed that DDT mimics estrogen in MCF-7 cells [45]. Bisphenol A (BPA) is the monomeric component of polycarbonate plastics used in many consumer products and has estrogenic activity in MCF-7 cells in vitro [46]. BPA has been shown to induce hyperprolactinemia in estrogen-sensitive Fisher 344 rats by increasing PRL gene expression in the pituitary gland. Furthermore, BPA exerts its influence at the PRL gene transcription level by activating the estrogen response element (ERE) in pituitary cells [47]. Although there is a body of evidence to support xenoestrogens as a possible public health hazard, the carcinogenic potency of these compounds is much lower than that of estradiol [48].

Cell Cycle, Genomic Instability, and Tumorigenesis in Lactotropes in Estrogen-Treated Rats

Factors Regulating Estrogen Mitogenic Action

A number of factors are both estrogen-dependent and affect lactotropic proliferation, differentiation, and/or transformation. However, the relatedness of these factors in the mediation of estrogen action on cell functions is not well understood. Some such estrogen-regulated factors include epidermal growth factor (EGF), platelet-derived growth factor (PDGF), insulin-like growth factor (IGF)-1, transforming growth factor-α (TGF-α), basic fibroblast growth factor (bFGF), fibroblast growth factor-4 (FGF-4), interleukin-2 (IL-2) and IL-6, nerve growth factor (NGF), and transforming growth factor-β (TGF-β). A stimulatory action for EGF, TGF-α, PDGF, IGF-1, and IGF-2 is shown on mesenchymal cell growth. EGF, TGF-α, IGF-1 and IGF-2 also stimulate epithelial cell growth [49]. During the developmental period, estrogen can stimulate the growth of all cell types; however, in the adult, estrogen action is restricted primarily to epithelial cells [50]. It is not known why such a dichotomy exists in the adult. It could be due to expression or lack of expression of inhibitory or stimulatory factors that interact with estrogen. Estrogen's action on epithelial cells often depends on the presence of mesenchymal cells [51]. This is yet another way in which estrogen-varied actions could occur through common receptors. The response to estrogen would depend not only on the cell that displays proliferation or differentiation but also on neighboring cells. This is supported by the fact that estrogen actions in some cases can be displayed only in vivo or in specific culture conditions. Although the actions of many of these estrogen-responsive growth regulatory factors have been tested in tumor lactotropes (e.g. GH3 cells) or in mixed pituitary cell populations [52–58], the actions of most of these factors in primary lactotropes have not been tested or persistent hyperprolactinemia was observed in 16 alcoholic women during a 6-week treatment trial [32]. These patients reported daily alcohol intake of 170 g for a period between 2 and 16 years period but had no clinical evidence of alcoholic liver cirrhosis. In a study reported by Japanese scientists, 22 of 23 women admitted for alcoholism treatment had PRL levels above normal that ranged between 27 and 184 ng/ml. These women reported drinking an average of 84.1 g of alcohol each day for at least 7 years. None of these patients showed liver cirrhosis, but 10 had hepatitis and the rest had fatty liver [33]. Studies conducted in a Massachusetts general hospital in the US reported hyperprolactinemia (22–87 ng/ml) in 6 of 12 alcohol-dependent women who had a history of drinking 75–247 g of alcohol per day for a minimum period of 7 years [34]. Alcohol-induced hyperprolactinemia is also reported in healthy, well-nourished women during residence on a clinical research ward for 35 days [35]. Sixty percent of women in the heavy drinker category (blood alcohol level 109–199 mg/dl) and 50% of moderate drinkers (blood alcohol levels 48–87 mg/dl) showed elevated PRL levels, and many of these drinkers had elevated PRL several days after cessation of drinking. Alcohol-induced hyperprolactinemia was also evident in 66 postmenopausal women [36]. The increase in PRL in these patients, however, was associated with increased androgen conversion to estradiol, possibly due to liver cirrhosis. Alcoholic men also showed elevated plasma levels of PRL [37–39]. Male alcoholic patients frequently show evidence of feminization that is manifested by gynecomastia, spider angiomata, palmar erythema and changes in body hair patterns [10]: a potential role for PRL and estradiol in the pathogenesis for the observed feminization. Thus, it appears that chronic alcohol administration in humans causes increased estrogen production and PRL elevation.

Alcohol-induced hyperprolactinemia has also been demonstrated in nonhuman primates [40]. Studies conducted in macaque female monkeys showed that in some, but not all, of the monkeys, the PRL levels were elevated after chronic self-administration of high doses of alcohol (3.4 g/kg/day). Interestingly, in one of these monkeys, immunocytochemical examination of the pituitary gland showed apparent pituitary hyperplasia. In laboratory rats, chronic ethanol administration increases plasma PRL levels and pituitary weight in cyclic female rats and ovariectomized rats as well as potentiating estrogen mitogenic effects in ovariectomized female rats [42, 43]. Therefore, clinical as well as animal data suggest that ethanol consumption is a positive risk factor for prolactinomas and hyperprolactinemia, possibly due to ethanol's ability to increase endogenous estrogenic activity.

Xenoestrogen-Treated Rats

Xenoestrogens are environmentally occurring chemicals that mimic the action of estrogen by binding to estrogen receptors [44]. Examples of xenoestrogens confirmed. Also, their actions on the lactotropes in the presence of pituitary mesenchymal variant folliculo-stellate cells have not been tested. Additionally, overexpression of many of these factors using a transgenic approach have resulted only in hyperplasia, and they resulted in adenoma formation only in old age and/or only in females, suggesting that endogenous estrogens are a requirement [59–61]. Therefore, the role of these growth factors in the mechanisms leading to tumorigenesis could not be critically evaluated. However, substantial evidence is now available to show that a disruption of the TGF-β-regulated cell growth balance is critical in the development of prolactinoma in the pituitary.

Cell Growth Balance by TGF-β Isoforms

The TGF-β family of peptides includes many related factors that exert numerous actions. In mammals, TGF-β is found in three highly homologous isoforms that exert their actions via heteromeric complexes of type I and type II receptors. TGF-β1 is the most well characterized of the three TGF-βs (TGF-β1, TGF-β2 and TGF-β3) [62]. These peptide growth factors function in cell proliferation, differentiation, embryogenesis, tissue repair, and cell adhesion. In most epithelial cell types, TGF-β1 acts as a negative growth regulator [63]. In mesenchymal cells, TGF-β1 has been shown to either increase or decrease cell proliferation. TGF-β1 can block the mitogenic actions of PDGF, FGF and EGF in certain cell types [63, 64]. Given the important regulatory role of TGF-β1 in controlling cell proliferation, it follows that a reduction in TGF-β1 could have implications on cell transformation.

TGF-β1 is produced in and secreted by pituitary lactotropes [65]. TGF-β1 is a potent inhibitor of estrogen-induced lactotropic proliferation and PRL secretion [65, 66]. Intrapituitary administration of TGF-β1 resulted in inhibited PRL release, decreased pituitary weight, and reduced DNA synthesis in estrogen-treated Fischer 344 rats. The inhibitory response to TGF-β1 is reduced in anterior pituitary cell lines, including the prolactin-secreting PR1 cell line and the GH and prolactin-secreting GH3 cell line. Both of these cell lines show low levels of TGF-β1 and its type II receptor (TβR-II) mRNA [67]. Lactotropes exposed to estrogen express reduced levels of TβR-II and TGF-β1 [68, 69]. TGF-β1 has been shown to interfere with the mitogenic actions of several growth-promoting agents including FGF, EGF and IL-1 [64]. This leads to the possibility that the mitogenic action of estrogen could involve its ability to decrease TGF-β1 levels, thereby allowing an increase in the levels of positive growth regulators.

Differential Expression of TGF-β Isoforms

In addition to TGF-β1, lactotropes also produce and secrete TGF-β3 [70]. Interestingly, however, TGF-β1 expression is reduced in lactotropes after estradiol exposure, while TGF-β3 expression is increased after the steroid treatment. Though the finding that a stimulus causes differential expression of two TGF-β isoforms in a single cell type is extremely novel, it is not unexplainable. Each of the TGF-β isoforms is the product of a separate gene, each with distinct regulatory regions. Furthermore, it has been shown that the TGF-β1 promoter has no TATAA box and is regulated principally by AP-1 sites, whereas both the TGF-β2 and TGF-β3 promoters have TATAA boxes as well as AP-2 sites and cAMP-responsive elements [71]. The AP-1 sites were previously thought to be specific for *fos/jun* transcription factors; however, it is now known that AP-1 sites can serve as nonclassical estrogen responsive elements [72]. Recently, Yang et al. [74] have demonstrated that estradiol selectively stimulates transcription of TGF-β3 in bone cells. This effect of estradiol does not occur via a classical estrogen-responsive element (ERE), but rather via an alternative estrogen-sensitive sequence. This raises the possibility that the pituitary may be a site where this alternative response element on the TGF-β3 gene is activated.

Opposing Actions of Two TGF-β Isoforms

In the lactotropes, the production of TGF-β1 and TGF-β3 not only differ under estrogenic condition, but these two peptides also act differently in the presence of estrogen. Unlike TGF-β1, TGF-β3 stimulates lactotropic cell proliferation in the presence of estrogen. In the absence of estrogen, TGF-β3 has a minimal effect on lactotropic cell proliferation [31]. Another interesting observation is that while TGF-β1 neutralizing antibody failed to prevent estradiol's mitogenic action, it potentiated the cell-proliferating action of TGF-β3 [70]. On the other hand, TGF-β3-neutralizing antibody inhibited lactotropic cell proliferation by estradiol. These data indicated that, unlike many other tissues, TGF-β1 and TGF-β3 have opposite actions on lactotropic cell proliferation. Furthermore, simultaneous suppression of the growth inhibitory TGF-β1 and activation of the growth stimulatory TGF-β3 may be a mechanism that estradiol uses to induce lactotropic cell proliferation.

Role of TGF-β Receptors in Mediation of TGF-β Isoform Actions

The literature suggests that the different isoforms of TGF-β act using similar receptor complexes (TGF-β type I and TGF-β type II receptors) to control growth response [62, 74]. However, the affinity for each ligand may vary [75]. The opposing actions of TGF-β1 and TGF-β3 on estrogen-induced lactotropic proliferation raise the questions of how the two isoforms could act differentially at the same receptor site, and of why the TGF-β3 that lactotropes produce would not be self-inhibitory. One explanation for TGF-β3's inability to be self-inhibitory is that the TGF-β receptors on lactotropes have a low affinity for TGF-β3 [76]. Another possibility is that estrogen suppresses TGF-β type II receptor that prevents the effect of TGF-β on lactotropes [67]. It has been shown that TGF-β1 inhibits lactotropic cell proliferation and that this response is mediated largely via the TGF-β type II receptors [67]. The TGF-β type I receptor in the pituitary has been implicated to be more involved in transcription responses evoked by TGF-β [76]. Recently, it has been demonstrated that TGF-β type II heterozygous 'knock-out' mice rapidly develop prolactinomas in response to estradiol exposure, unlike control mice [77]. A rat pituitary derived PR1 cell line, which has reduced levels of type II receptors [68, 76], displays very little or no growth-inhibitory response to TGF-β1 [67]. However, in the presence of TGF-β type I receptors and very low expression of TGF-β type II receptors, TGF-β1 inhibits PRL secretion in PR1 cells [76]. A similar scenario may exist in the case of FS cells, where the action of TGF-β3 on gene transcription (production of b-FGF) predominates because of the normal expression of the type I receptor even though there is a low expression of the TGF-β type II receptor under estrogenic conditions.

Communication between Folliculo-Stellate Cells and Lactotropes

In several tissues, it appears that the growth of cells depends not only on the mitogenic stimulus but also on cell-to-cell communication. For example, in uterine, ovarian, and mammary tissues, communication with mesenchymal cells facilitates the growth of epithelial cells [56]. Recently, it has been shown that FS cells regulate estradiol's cell-proliferating action on lactotropes [70]. FS cells are characterized by their stellate shape and long cytoplasmic processes, and they are largely devoid of secretory granules. S-100 immunoreactivity has proven to be a reliable marker of FS cells [78]. It has been suggested that FS cells perform several supportive functions, including regulation of phagocytosis [79], secretion of angiogenic factors [80], growth factors [31] and cytokines [81], as well as tropic and stem cell functions [82]. FS cells seem to be targeted by estrogen because they rapidly increase c-fos expression [83] in response to estrogen.

FS cells have been shown to participate in estrogen's mitogenic action on lactotropes [31]. Estradiol induces TGF-β3 release from lactotropes. TGF-β3 then acts on FS cells to release bFGF, which acts on lactotropes to increase cell proliferation. Thus, FS cells mediate estradiol's mitogenic action on lactotropes by releasing bFGF. FS cells also appear to determine the susceptibility of lactotropes to estradiol [84]. When co-transplanted with Fischer-344 pituitaries under the kidney capsule or co-cultured with Fischer-344-derived lactotropes in vitro, FS cells derived from Fischer-344 rats increased estradiol's mitogenic action. They also increased estradiol's growth-promoting action on Sprague-Dawley-derived lactotropes (which are normally minimally responsive to estrogen) in primary cultures. However, Sprague-Dawley-derived FS cells failed to increase estrogen's action on Fischer-344- or on Sprague Dawley-derived lactotropes. The levels of bFGF production and secretion by TGF-β3 and estradiol were much higher in Fischer-344-derived FS cells than in Sprague-Dawley-derived FS cells. However, the growth response of lactotropes to bFGF was similar in both strains. These data suggest that cell-cell interaction between FS cells and lactotropes is important in the regulation of estradiol's mitogenic action on lactotropes and on the lactotropes' susceptibility to the steroid.

Gap-Junctional Communication

In the rat pituitary, formation of gap junctions between FS cells has been demonstrated using electron microscopic methods [85, 86]. It has also been shown that FS cells communicate with other FS cells and with endocrine cells, including prolactin-secreting lactotropes [87]. The gap-junctional number in the anterior pituitary varied during the estrous cycle; it was highest in proestrous and in estrous and increased during middle to late pregnancy in rats [85]. This suggests that gonadal factors such as estradiol may regulate the rate of occurrence of gap junction in the anterior pituitary gland. Gap junctions connect the cytoplasmic domains of contacting cells to allow metabolic and ionic exchange between them [87]. Gap junctions consist of channel proteins called connexins, a family of proteins which, upon phosphorylation, increase permeabilization of the cells to molecules that are permeable through gap junctions [87]. Among the various types of connexins, connexin 43 (Cx43) is known to express in astrocytes and the pituitary [88]. It has recently been shown that there were no changes in dye transfer or Cx43 in the FS cells following estradiol [89], indicating that the steroid had no direct effect on the molecular transfer across gap junctions in the FS cells. Since FS cells were shown to be activated by the TGF-β3 secreted by the lactotropes [31], increased gap-junctional communication after TGF-β3 may also underlie the modulatory role of FS cells in pituitary function and pathology. For instance, increased gap-junctional communication could lead to more FS cells being activated by lactotropes that could then lead to more bFGF release by the FS cells that, in turn, could further increase lactotropic cell growth.

Mitogen-Activated Protein Kinase Signaling

The MAP kinase p44/42 has also been shown to be regulated by estrogen [90]. The genomic effects of estrogen occur primarily through interaction with estrogen receptors [91]. The nongenomic actions of estrogen include activation of Ras [92], which in turn activates MAP kinase p44/42, Raf-1 [93], protein kinase C [94], protein kinase A [95], and Maxi-K channels [96]. These actions of estrogen may be mediated by a plasma membrane-associated estrogen receptor or by nonclassical estrogen receptors [97]. Studies of GH3 cells revealed that anti-estrogen receptor antibodies, directed against the hinge region of the estrogen receptor, blocked rapid activation of MAP kinase p44/42 and estradiol-induced PRL secretion [98]. In some cell systems, estradiol has been shown to activate a rapid MAP kinase p44/42 response within 3–5 min of treatment, as well as a delayed sustained response [91]. For rapid effects of estrogen, the receptors must be located in the plasma membrane. Delayed effects could be attributed to classical nuclear receptor and genomic action, as well to the paracrine action of estradiol through the secretion of growth factors [90]. Increasing evidence in recent years supports the existence of an alternative receptor that is both genetically and pharmacologically different from a classical estrogen receptor [97] and insensitive to the estrogen receptor blocker ICI 182, 780 [99]. A slow but persistent activation of MAP kinase p44/42 was observed after treatment with estradiol and TGF-β3 to induce bFGF production from FS cells [100]. The estrogen receptor blocker ICI 182,780 did not block bFGF levels in FS cells induced by estradiol or by TGF-β3 and estradiol, suggesting the possibility of a classical estrogen receptor-independent mechanism to increase bFGF production. Recent studies also provide evidence that TGF-β and estradiol interact to increase bFGF production and release from FS cells via the estradiol receptor-independent PKC-Ras-MEK-MAP kinase p44/42 signaling pathway [100].

Involvement of PKC-Independent MAP Kinase p44/42 Signaling in Mediation of bFGF Action on Proliferation of Lactotropes. bFGF has been shown to cause proliferation of pituitary lactotropic cells [31, 101]. bFGF is a member of a large group of FGF peptides that control multiple cellular processes, including cell migration, survival, proliferation and differentiation [102]. FGF peptides mediate their biological responses by binding to and activating a family of receptor tyrosine kinases (RTKs) designated as fibroblast growth factor receptors (FGFRs) [103]. The peptide bFGF possesses a dual receptor system consisting of tyrosine kinase FGFRs and heparin sulphate proteoglycans. It has been found that, in most cases, both of these components must be present in order for bFGF to trigger proliferation [104]. Binding of bFGF to the receptor is known to cause receptor dimerization and activation of the cytoplasmic domain. These RTKs subsequently activate multiple signaling pathways, including protein kinase C (PKC), Src kinases, Ras, and MAP kinase p44/42 [105, 106]. bFGF induces proliferation of lactotropic cells by activating MAP kinase p44/42 phosphorylation. Compelling evidence has recently been provided to show that bFGF increases lactotropic cell proliferation via PKC-independent Src-Ras-MEK-MAP kinase p44/42 signaling [107].

Dopamine D2 Receptors and TGF-β Type II Receptor Interaction

Dopamine plays a key role in maintaining the normal function of lactotropes in the pituitary gland. Abnormalities in dopamine receptors and dopamine transporter function lead to hyperplasia of lactotropes [28, 108–110]. It is well known that estradiol inhibits dopamine release from the hypothalamus and downregulates dopamine D2 receptor activity in lactotropes [15, 84, 111]. The steroid decreases the potency of dopamine to inhibit PRL release by uncoupling the D2 receptors with Gαi/αo. The uncoupling of the D2 receptor might be the result of the phosphorylation of Gαi/αo [112] and downregulation of Gαi/αo protein synthesis [113]. Estrogen also regulates the expression of a splicing factor thus favoring the production of the D2L [114]. The D2 receptor is a 7-transmembrane segment protein with a long third intracellular loop and a short intracellular C-terminus. The sixth exon of the D2 receptor gene is often excluded in the mature transcript, resulting in a short (29 amino acids shorter) isoform (D2S). Estradiol strongly favors the expression of the long isoform (D2L) mRNA over the short isoform D2S [114], thus causing uncoupling to the G protein [115, 116]. It is interesting to note that ethanol, which increases lactotropic cell proliferation, also favors production of the D2L over the D2S isoform [84].

Treatment with dopamine agonists in some patients reverses the hyperprolactinemia [8]. This reversal supports the notion developed from experimental models that prolactinoma formation results from a disruption in dopamine function [15, 117]. Our laboratory has recently identified a novel signaling cascade for lactotrope growth control involving the D2S, TGF-β1 and its type II receptor [118]. It has recently been shown that transgenic mice, overexpressing D2S but not D2L, show pituitary hypoplasia; the D2S overexpressing mice also showed increased phosphorylated MAPK [28]. The MAPK system has also been shown to be involved in TGF-β1-activated signaling in various cell types [100, 119]. Hence, the possibility that the MAPK pathway is involved in dopamine-TGF-β1 interaction mechanisms needs to be investigated.

Estrogen is known to reduce both the levels of dopamine and of dopamine D2 receptor activity in lactotropes [15, 84, 120]. By inhibiting dopamine, estrogen could inhibit TGF-β1 and its receptor expression. Indeed, this association has been observed following long-term estrogen treatment [68]. In addition, we have observed that GH3 cells that have reduced functional dopamine D2 receptors have a low TGF-β1 response and low TGF-β1 production and reduced TβRII levels [68]. We have also shown that PR1 cells, which did not respond to TGF-β1 and did not express detectable TGF-β1 or TβRII, showed a lack of dopamine response and dopamine receptor binding. Hence, we have proposed that during sustained exposure, estrogen cancels the inhibitory effect of dopamine and thereby downregulates the inhibitory effect of TGF-β1 on cell proliferation. This

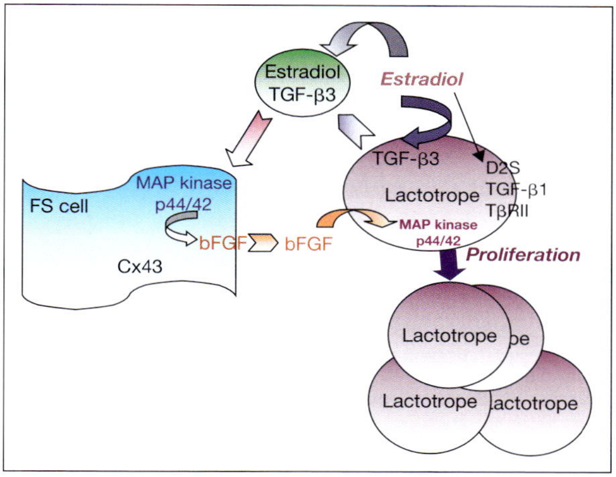

Fig. 1. Diagram summarizing the proposed roles of the FS cell in mediation of estradiol's actions on lactotropic cell proliferation. ➡ = Stimulation; → = inhibition.

may cause an alteration in the balance between positive and negative regulators of cell growth, resulting in abnormal lactotropic cell proliferation.

Conclusion

Lactotrope sensitivity to the mitogenic action of estradiol has important implications in normal physiology. During pregnancy, estrogen levels are elevated, which causes an increased number of lactotropes and increased plasma PRL, which is necessary for the initiation of milk production. When the level of estrogen returns to basal levels, the mitogenesis of lactotropes ceases. If estradiol levels remain elevated for prolonged periods, tumor formation can result. Although prolactinomas are quite common in the human population and are readily induced in animal models, our knowledge of the formation of these tumors is limited. The data presented in this review indicate that estradiol exposure may lead to prolactinoma formation via orchestrated events involving dopamine D2 receptors, TGF-β isoforms and their receptors, as well as factors secondary to TGF-β action. These data demonstrate that estradiol suppresses the dopamine D2 receptor's splice variant D2S to reduce TGF-β1-TβRII inhibitory signaling in lactotropic cell proliferation. An estradiol inhibitory action on TGF-β1 and or TβRII expression leads to loss of growth-inhibitory control of TGF-β1 in the lactotropes. The steroid also increases TGF-β3 production and secretion from lactotropes. The secreted TGF-β3 is transported to the neighboring FS cells, where it acts to induce the release of bFGF. The FS cell-derived b-FGF stimulates lactotropic cell proliferation as lactotropes escape from TGF-β1 growth inhibition and are activated by estradiol to express FGF receptors. We presented data to show that MAP kinase p44/42 phosphorylation via the activation of Src kinases is required for cell growth. We postulate that the loss of TGF-β1 growth-inhibitory control may be a contributing factor for lactotropic cell transformation. Taken together, these studies further our understanding not only of TGF-β action in the anterior pituitary but also of the etiology of estradiol-induced prolactinomas (fig. 1). The identification of TGF-β1 mediation of dopamine action also provides a novel possibility to consider the TGF-β1-TβRII signaling as a molecular target for treating prolactinomas.

References

1 Asa SL, Ezzat S: The cytogenesis and pathogenesis of pituitary adenomas. Endocr Rev 1998;19:798–827.
2 Levy A, Lightman S: Molecular defects in the pathogenesis of pituitary tumors. Front Neuroendocrinol 2003;24:94–127.
3 Ishibashi M, Yamaji T: Effects of hypophysiotropic factors on growth hormone and prolactin secretion from somatotroph adenomas in culture. J Clin Endocrinol Metab 1985;60:985–988.
4 Hardy J: Ten years after the recognition of pituitary microadenomas; in Faglia G, Giovanelli MA, MacLeod RM (eds): Pituitary Microadenomas. New York, Academic Press, 1980, pp 7–14.
5 Miyai K, Ichihoza K, Kondo K, Mori S: Asymptomatic hyperprolactinoma in the general population mass screening by paired assays of serum prolactin. Clin Endocrinol 1986;25:549–552.
6 Kelley WN (ed): Text Book of Internal Medicine. New York, Lippincott, 1996, pp 1–1983.
7 MacLeod RM, Scapagnini U, Thorner MO: Pituitary Microadenomas. New York, Academic Press, 1980, pp 683–855.
8 Tonner D, Schlechte J: Contemporary therapy of prolactin-secreting adenomas. Am J Med Sci 1993;306:395–397.
9 Powell M, Lightman SL: The Management of Pituitary Tumors: A Handbook. Edinburgh, Churchill-Livingstone, 1996.
10 Van Thiel DH: Feminization of chronic alcoholic men: a formulation. Yale J Biol Med 1979;52:219–225.
11 Sarkar DK, Gottschall PE, Meites J: Relation of the neuroendocrine system to development of prolactin secretion pituitary tumors; in Meites J (ed): Neuroendocrinology of Aging. New York, Plenum Press, 1983, pp 353–376.
12 Sadoul JL, Thyss A, Freychet P: Invasive mixed growth hormone/prolactin secreting pituitary tumor: complete shrinking by octreotide withdrawal. Acta Endocrinol 1992;126:179–183.
13 DeNicola AF, Von Lawzewifsch IS, Kaplan E, Libertum C: Biochemical and ultrastructural studies of estrogen-induced pituitary tumors in F344 rats. J Natl Cancer Inst 1978;61:753–763.
14 Wicklund JA, Wertz N, Gorski J: A comparison of estrogen effect on uterine and pituitary growth and prolactin synthesis in F-344 and Holtzman rats. Endocrinology 1981;109:1700–1707.
15 Sarkar DK, Gottschall PE, Meites J: Damage to hypothalamic dopaminergic neurons is associated with development of prolactin-secreting tumors. Science 1982;218:684–686.
16 Banerjee SK, De A, Sarkar DK: Colocalization of prolactin and proliferating cell nuclear antigen in the anterior pituitary during estrogen-induced pituitary tumors. Cancer Lett 1994;87:139–144.
17 Shull JD, Spady TJ, Snyder MC, Johansson SL, Pennington KL: Ovary-intact, but not ovariectomized female ACI rats treated with 17β-estradiol rapidly develop mammary carcinoma. Carcinogenesis 1997;18:1595–1601.

18 Lloyd RV: Estrogen-induced hyperplasia and neoplasia in the rat anterior pituitary gland. Am J Pathol 1983;113:198–206.
19 Gooren LJ, Assies J, Asscheman H, de Slegte R, van Kessel H: Estrogen-induced prolactinoma in a man. J Clin Endocrinol Metab 1988;66:444–446.
20 Garcia MM, Kapcala LP: Growth of a microprolactinoma to a macroprolactinoma during estrogen therapy. J Endocrinol Invest 1995;18:450–455.
21 Carol W, Lauterbach H, Klinger G, Unger A, Michels W: Prolactin stimulation using the metoclopramide test in females taking oral contraceptives. Zentralbl Gynakol 1988;110:1515–1521.
22 Shy KK, McTiernan AM, Daling JR, Weiss NS: Oral contraceptives use and the occurrence of pituitary prolactinoma. J Am Med Assoc 1983;249:2204–2207.
23 Luciano AA, Sherman BM, Chapler FK, Hauser KS, Wallace RB: Hyperprolactinemia and contraception: a prospective study. Obstet Gynecol 1985;65:506–511.
24 Holmgren U, Bergstrand G, Hagenfeldt K, Werner S: Women with prolactinoma-effect of pregnancy and lactation on serum prolactin and on tumor growth. Acta Endocrinol (Copenh) 1986;111: 452–459.
25 Gíomez F, Reyes F, Fairman C: Nonpuerperal galactorrhea and hyperprolactinemia: clinical findings, endocrine features and therapeutic responses in 56 cases. Am J Med 1977;62:648–660.
26 Sherman B, Schlechte M, Halmi JNS, Chapler FK, Harris CE, Duello TM, Van Gilder J, Granner DK: Pathogenesis of prolactin-secreting pituitary adenomas. Lancet 1978;ii:1019–1021.
27 Chang RJ, Keye WR, Young JR, Willson CB, Jaffe RB: Detection, evaluation and treatment of pituitary microadenomas in patients with galactorrhea and amenorrhea. Am J Obstet Gynecol 1977;128:356–363.
28 Laccarino C, Samad TK, Mathis C, Kercret H, Picetti R, Borrelli E: Control of lactotrope proliferation by dopamine: essential role of signaling through D2 receptors ERKs. Proc Soc Nat Acad Sci USA 2002;99:14530–14535.
29 Oomizu S, Chaturvedi K, Sarkar DK: Folliculostellate cells determine the susceptibility of lactotropes to estradiol's mitogenic action. Endocrinology 2004;145:1473–1480.
30 Schechter JE, Pattison A, Pattison T: Development of the vasculature of the anterior pituitary: ontogeny of basic fibroblast growth factor. Dev Dyn 1993;197:81–93.
31 Hentges S, Boyadjieva N, Sarkar DK: Transforming growth factor-beta3 stimulates lactotrope cell growth by increasing basic fibroblast growth factor from folliculostellate cells. Endocrinology 2000;141:859–867.
32 Välimäki M, Pelkonen R, Harkonen M, Tuomala P, Koistinen P, Roine R, Ylikahri R: Pituitary-gonadal hormones and adrenal androgens in non-cirrhotic female alcoholics after cessation of alcohol intake. Eur J Clin Invest 1990;20:177–181.
33 Seki M, Yoshida K, Okamura Y: A study on hyperprolactinemia in female patients with alcoholics. Jpn J Alcohol Drug Depend 1991:26:49–59.
34 Teoh SK, Lex BW, Mendelson JH, Mello NK, Cochin J: Hyperprolactinemia and macrocytosis in women with alcohol and polysubstance dependence. J Stud Alcohol 1992;53:176.
35 Mendelson JH, Mello NK: Chronic alcohol effects on anterior pituitary and ovarian hormones in healthy women. J Pharmacol Exp Ther 1988;245:407–412.
36 Gavaler JS: Aging and alcohol: the hormonal status of postmenopausal women; in Sarkar DK, Barnes C (eds): Reproductive Neuroendocrinology of Aging and Drug Abuse. Boca Raton, CRC Press, 1994, pp 365–378.
37 Ida Y, Isujimararu S, Nakamaura K, Shirrao I, Mukasa H, Egami H, Nakazawa Y: Effects of acute and repeated alcohol ingestion on hypothalamic-pituitary-gonadal and hypothalamic-pituitary-adrenal functioning in normal males. Drug Alcohol Depend 1992;31:57–64.
38 Marchesi C, De Risio C, Campanini G, Majgini C, Piazza P, Grassi M, Chiodera P, Criro V: TRH test in alcoholics: relationship of the endocrine results with neuroendocrinological and neuropsychological findings. Alcohol Alcohol 1992;27:531–537.
39 Soyka M, Gorig E, Naber D: Serum prolactin increase induced by ethanol-a dose dependent effect not related to stress. Psychoneuroendocrinology 1991;16:441–446.
40 Kornet M, Goosen C, Thyssen JH, Van Ree JM: Endocrine profile during acquisition of free-choice alcohol drinking in rhesus monkeys: treatment with desglycinamide-(Arg 8)-vasopressin. Alcohol Alcohol 1991;27:403–410.

41 Mello NK, Mendelson JH, Bree MP, Skupny A: Alcohol effects on naloxone-stimulated luteinizing hormone, follicle-stimulating hormone and prolactin plasma levels in female Rhesus monkeys. J Pharmacol Exp Ther 1988;245:895–904.
42 Sanchis R, Esquifino A, Guerri C: Chronic ethanol intake modifies estrous cyclicity and alters prolactin and LH levels. Pharmacol Biochem Behav 1985;23:221–224.
43 De A, Boyadjieva N, Pastorcic M, Sarkar DK: Potentiation of estrogen's mitogenic effect on the pituitary gland by alcohol consumption. International J Oncol 1995;7:643–648.
44 Stone R: Environmental estrogens stir debate. Science 1994;265:308–310.
45 Dees C, Asakari M, Foster JS, Ahamed S, Wimalasena J: DDT mimics estradiol stimulation of breast cancer cells to enter cell cycle. Mol Carcinogenesis 1997;18:107–114.
46 Krishnan AV, Stathis P, Permuth SF, Tokes L, Feldman D: Bisphenol A: an estrogenic substance released from polycarbonate flask during autoclaving. Endocrinology 1993;132:2279–2286.
47 Steinmetz R, Brown NG, Allen DL, Bigsby RM, Ben-Jonathan N: The environmental estrogen bisphenol A stimulates prolactin release in vitro and in vivo. Endocrinology 1997;138: 1780–1786.
48 Hunter DJ, Hankinson SE, Laden F: Plasma organochlorine levels and the risk of breast cancer. N Engl J Med 1997;337:1253–1258.
49 Goustin AS, Leof EB, Shipley GD, Moses HL: Growth factors and cancer. Cancer Res 1986;46: 1015–1029.
50 Clark BF: The effect of oestrogen and progesterone on uterine cell division and epithelial morphology in spayed, adrenalectomized rats. J Endocrinol 1971;50:527–528.
51 Parrot JA, Vigne JLB, Chu Z, Skinner MK: Mesenchymal-epithelial interactions in the ovaries follicle involve keratinocyte and hepatocytes growth factor production by theca cells and their action on granular cells. Endocrinology 1994;135:569–575.
52 Simon I, Huttner A, Said J, Spirina OM, Melmed S: Heparin-binding secretory transforming gene (hst) facilitates rat lactotrope cell tumorigenesis and induces prolactin gene transcription. J Clin Invest 1996;97:187–195.
53 Atkin SL, Landolt AM, Foy P, Jeffreys RV, Hipkin L, White MC: Effects of insulin-like growth factor-I on growth hormone and prolactin secretion and cell proliferation of human somatotrophinomas and prolactinomas in vitro. Clin Endocrinol (Oxf) 1994;41:503–509.
54 Sullivan NJ, Tashjian AH Jr: Platelet-derived growth factor selectively decreases prolactin production in pituitary cells in culture. Endocrinology 1983;113:639–645.
55 Borrelli E, Sawchenko PE, Evans RM: Pituitary hyperplasia induced by ectopic expression of nerve growth factor. Proc Natl Acad Sci USA 1992;89:2764–2768.
56 Tichomirowa M, Theodoropoulou M, Lohrer P, Schaaf L, Losa M, Uhl E, Lange M, Arzt E, Stalla GK, Renner U: Bacterial endotoxin (lipopolysaccharide) stimulates interleukin-6 production and inhibits growth of pituitary tumour cells expressing the toll-like receptor 4. J Neuroendocrinol 2005;17:152–160.
57 Fedele M, Pentimalli F, Baldassarre G, Battista S, Klein-Szanto AJ, Kenyon L, Visone R, De Martino I, Ciarmiello A, Arra C, Viglietto G, Croce CM, Fusco A: Transgenic mice overexpressing the wild-type form of the HMGA1 gene develop mixed growth hormone/prolactin cell pituitary adenomas and natural killer cell lymphomas. Oncogene 2005;24:3427–3435.
58 Onofri C, Carbia Nagashima A, Schaaf L, Feirer M, Lohrer P, Stummer W, Berner S, Chervin A, Goldberg V, Stalla GK, Renner U, Arzt E: Estradiol stimulates vascular endothelial growth factor and interleukin-6 in human lactotroph and lactosomatotroph pituitary adenomas. Exp Clin Endocrinol Diabetes 2004;112:18–23.
59 McAndrew J, Paterson AJ, Asa SL, McCarthy KJ, Kudlow JE: Targeting of transforming growth factor-alpha expression to pituitary lactotrophs in transgenic mice results in selective lactotroph proliferation and adenomas. Endocrinology 1995:136:4479–4488.
60 Cruz-Soto ME, Scheiber MD, Gregerson KA, Boivin GP, Horseman ND: Pituitary tumorigenesis in prolactin gene-disrupted mice. Endocrinology 2002;143:4429–4436.
61 Hentges ST, Low MJ: Ovarian dependence for pituitary tumorigenesis in D2 dopamine receptor-deficient mice. Endocrinology 2002;143:4536–4543.
62 Massagué J: Receptors for the TGF-β family. Cell 1992;69:1067–1070.
63 Moses HL, Young EY, Pietenpol JA: Regulation of epithelial proliferation by TGF-beta. Cell 1990;63:245–247.

64 Sporn MB, Roberts AB: Autocrine growth factors and cancer. Nature 1985;313:745–747.
65 Sarkar DK, Kim KH, Minami S: Transforming growth factor-β1 mRNA and protein expression in the pituitary gland and its action on PRL secretion and lactotropic growth. Mol Endocrinol 1992;6:1825–1833.
66 Minami S, Sarkar DK: Transforming growth factor-β1 inhibits prolactin secretion and lactotropic cell proliferation in the pituitary of estrogen-treated Fischer 344 rats. Neurochem Int 1997;30: 499–506.
67 De A, Morgan TE, Speth RC, Boyadjieva N, Pastorcic M, Sarka DK: Pituitary lactotrope expresses TGF-β type II receptor mRNA and protein and contains [^{125}I]TGF-β1-binding sites. J Endocrinol 1996;149:19–27.
68 Pastorcic M, De A, Boyadjieva N, Vale W, Sarkar DK: Reduction in the expression and action of transforming growth factor β1 on lactotropes during estrogen-induced tumorigenesis. Cancer Res 1995;55:4892–4898.
69 Pastorcic M, Sarkar DK: Downregulation of TGF-β1 gene expression in anterior pituitary cells treated with forskolin. Cytokine 1997;9:106–111.
70 Hentges S, Pastorcic M, De A, Boyadjieva N, Sarkar DK: Opposing actions of two transforming growth factor-beta isoforms on pituitary lactotropic cell proliferation. Endocrinology 2000;141:1528–1535.
71 Roberts AB, Kim SJ, Noma T, Glick AB, Lafyatis R, Lechleider R, Jakowlew SB, Geiser A, O'Reilly MA, Danielpour D: Multiple forms of TGF-beta: distinct promoters and differential expression. Ciba Found Symp 1991;157:7–15.
72 Day RN, Koike S, Sakai M, Muramatsu M, Maurer RA: Both Pit-1 and the estrogen receptor are required for estrogen responsiveness of the rat prolactin gene. Mol Endocrinol 1990;4:1964–1971.
73 Yang NN, Venugopalan M, Sushant H, Glasebrook A: Identification of an estrogen response element activated by metabolites of 17β-estradiol and raloxifene. Science 1996;273:1222–1224.
74 Derynck R: TGF-β-receptor-mediated signaling. TIBS 1994;19:548–553.
75 Arrick BA, Korc M, Derynck R: Differential regulation of expression of three transforming growth factor beta species in human breast cancer cell lines by estradiol. Cancer Res 1990;50:299–303.
76 Sarkar DK, Hentges ST, De A, Reddy RHR: Hormonal control of pituitary prolactin-secreting tumors. Front Biosci 1998;3:D934–D943.
77 Shida N, Ikeda H, Takashi Y, Oshima M, Taketo MM, Miyoshi I: Estrogen-induced tumorigenesis in the pituitary gland of TGf-β(+/–) knockout mice. Biochim Biophys Acta 1998;1407:79–83.
78 Nakajima T, Yamaguchi H, Takahashi K: S 100 protein in folliculostellate cells of the rat pituitary anterior lobe. Brain Res 1980;191:523–531.
79 Gracia-Navarro F, Porter D, Garcia-Navarro S, Licht P: Immunocytochemical and ultrastructural study of the frog (*Rana pipiens*) pars distalis with special reference to folliculostellate cell function during in vitro superfusion. Cell Tissue Res 1989;256:623–630.
80 Gloddek J, Pagotto U, Paez Pereda M, Arzt E, Stalla GK, Renner U: Pituitary adenylate cyclase-activating polypeptide, interleukin-6 and glucocorticoids regulate the release of vascular endothelial growth factor in pituitary folliculostellate cells. J Endocrinol 1999;160:483–490.
81 Vankelecom H, Andries M, Billiau A, Denef C: Evidence that folliculostellate cells mediate the inhibitory effects of interferon-gamma on hormone secretion in rat anterior pituitary cell cultures. Endocrinology 1992;130:3537–3546.
82 Horvath E, Kovacs K: Folliculostellate cells of the human pituitary: a type of adult stem cell? Ultrastruct Pathol 2002;26:219–228.
83 Allen DL, Mitchner NA, Uveges TE, Nephew KP, Khan S, Ben-Jonathan N: Cell-specific induction of c-fos expression in the pituitary gland by estrogen. Endocrinology 1997;138:2128–2135.
84 Oomizu S, Boyadjieva N, Sarkar DK: Ethanol and estradiol modulate alternative splicing of dopamine D2 receptor mRNA and abolish the inhibitory action of bromocriptine on prolactin release from the pituitary gland. Alcohol Clin Exp Res 2003;27:975–980.
85 Soji T, Mabuchi Y, Kurono C, Herbert DC: Folliculo-stellate cells and intercellular communication within the rat anterior pituitary gland. Microsc Res Tech 1997;39:138–149.
86 Morand I, Fonlupt P, Guerrier A, Trouillas J, Calle A, Remy C, Rousset B, Munari-Silem Y: Cell-to-cell communication in the anterior pituitary: evidence for gap junction-mediated exchanges between endocrine cells and folliculostellate cells. Endocrinology 1996;137:3356–3367.

87 Bennett MV, Barrio LC, Bargiello TA, Spray DC, Hertzberg E, Saez JC: Gap junctions: new tools, new answers, new questions. Neuron 1997;6:305–320.
88 Evans WH, Martin PE: Gap junctions: structure and function. Mol Membr Biol 2002;19:121–136.
89 Kabir N, Chaturvedi C, Liu LS, Sarkar DK: TGF-β3 increases gap-junctional communication among folliculostellate cells to increase bFGF release. Endocrinology 2005;146:4054–4060.
90 Keshamouni VG, Mattingly RR, Reddy KB: Mechanism of 17-beta-estradiol-induced Erk1/2 activation in breast cancer cells: a role for HER2 and PKC-delta. J Biol Chem 2002;277:22558–22565.
91 van der Burg B, van Selm-Miltenburg AJ, de Laat SW, van Zoelen EJ: Direct effects of estrogen on c-fos and c-myc protooncogene expression and cellular proliferation in human breast cancer cells. Mol Cell Endocrinol 1989;64:223–228.
92 Weigel NL: Steroid hormone receptors and their regulation by phosphorylation. Biochem J 1996;319:657–667.
93 Reddy KB, Yee D, Hilsenbeck SG, Coffey RJ, Osborne CK: Inhibition of estrogen-induced breast cancer cell proliferation by reduction in autocrine transforming growth factor alpha expression. Cell Growth Differ 1994;5:1275–1282.
94 Arteaga CL, Coronado E, Osborne CK: Blockade of the epidermal growth factor receptor inhibits transforming growth factor alpha-induced but not estrogen-induced growth of hormone-dependent human breast cancer. Mol Endocrinol 1988;2:1064–1106.
95 Dickson RB, Lippman ME: Estrogenic regulation of growth and polypeptide growth factor secretion in human breast carcinoma. Endocr Rev 1987;8:29–34.
96 Korovkina VP, Brainard AM, Ismail P, Schmidt TJ, England SK: Estradiol binding to maxi-K channels induces their down-regulation via proteasomal degradaion. J Biol Chem 2004;272:1217–1223.
97 Nadal A, Ropero AB, Fuentes E, Soria B: The plasma membrane estrogen receptor: nuclear or unclear. Trends Pharmacol Sci 2001;22:597–599.
98 Watson CS, Norfleet AM, Pappas TC, Gametchu B: Rapid actions of estrogens in GH3/B6 pituitary tumor cells via a plasma membrane version of estrogen receptor-alpha. Steroids 1999;64: 5–13.
99 Gu Q, Korach KS, Moss RL: Rapid action of 17beta-estradiol on kainate-induced currents in hippocampal neurons lacking intracellular estrogen receptors. Endocrinology 1999;140:660–666.
100 Chaturvedi K, Sarkar DK: Involvement of PKC dependent p44/42 MAP kinase signaling pathway for cross-talk between estradiol and TGF-β3 in increasing bFGF in folliculostellate cells. Endocrinology 2004;145:706–715.
101 Prysor-Jones RA, Silverlight JJ, Jenkins JS: Oestradiol, vasoactive intestinal peptide and fibroblast growth factor in the growth of human pituitary tumour cells in vitro. J Endocrinol 1989;120:171–177.
102 Naski MC, Ornitz DM: FGF signaling in skeletal development. Front Biosci 1998;3:D781–D794.
103 Jaye M, Schlessinger J, Dionne CA: Fibroblast growth factor receptor tyrosine kinases: molecular analysis and signal transduction. Biochem Biophys Acta 1992;1135:185–199.
104 Delehedde M, Lyon M, Sergeant N, Rahmoune H, Fernig DG: Proteoglycans: pericellular and cell surface multireceptors that integrate external stimuli in the mammary gland. J Mammary Gland Biol Neoplasia 2001;6:253–273.
105 Mohammadi M, Dionne CA, Li W, Li N, Spivak T, Honegger AM, Jaye M, Schlessinger J: Point mutation in FGF receptor eliminates phosphatidylinositol hydrolysis without affecting mitogenesis. Nature 1992;358:681–684.
106 Klint P, Kanda S, Kloog Y, Claesson-Welsh L: Contribution of Src and Ras pathways in FGF-2 induced endothelial cell differentiation. Oncogene 1999;18:3354–3364.
107 Chaturvedi K, Sarkar DK: Mediation of basic fibroblast growth factor-induced lactotropic cell proliferation by Src-Ras-mitogen-activated protein kinase p44/42 signaling. Endocrinology 2005;146:1948–1955.
108 Kelly MA, Rubinstein M, Asa SL, Zhang G, Saez C, Bunzow JR, Allen RG, Hnasko R, Ben-Jonathan N, Grandy DK, Low MJ: Pituitary lactotroph hyperplasia and chronic hyperprolactinemia in dopamine D2 receptor-deficient mice. Neuron 1997;19:103–113.
109 Saiardi A, Bozzi Y, Baik JH, Borrelli E: Antiproliferative role of dopamine: loss of D2 receptors causes hormonal dysfunction and pituitary hyperplasia. Neuron 1997;19:115–126.
110 Bosse R, Fumagalli F, Jaber M, Giros B, Gainetdinov RR, Wetsel WC, Missale C, Caron MG: Anterior pituitary hypoplasia and dwarfism in mice lacking the dopamine transporter. Neuron 1997;19:127–138.

111 Ben-Jonathan N, Hnasko R: Dopamine as a prolactin (PRL) inhibitor. Endocr Rev 2001;22: 724–763.
112 Maus M, Homburger V, Bockaert J, Glowinski J, Premont J: Pretreatment of mouse striatal neurons in primary culture with 17β-estradiol enhances the pertussis toxin-catalyzed ADP-ribosylation of $G\alpha_{oi}$ protein subunits. J Neurochem 1990;55:1244–1251.
113 Livingstone JD, Lerant A, Freeman ME: Ovarian steroids modulate responsiveness to dopamine and expression of G-proteins in lactotropes. Neuroendocrinology 1998;68:172–179.
114 Guivarc'h D, Vincent J-D, Vernier P: Alternative splicing of the D2 dopamine receptor messenger ribonucleic acid is mediated by activated sex steroid receptors in the MMQ prolactin cell line. Endocrinology 1998;139:4213–4221.
115 Kobilka BK, Kobilka TS, Danil K, Regan JW, Caron MG, Lefkowitz RJ: Chimeric α2-, β2-adrenergic receptors: delineating of domaine involved in effector coupling and ligand binding specificity. Science 1988;240:1310–1316.
116 Guiramand J, Montmayeur J-P, Crealine J, Bhatia M, Borrelli E: Alternative splicing of the dopamine D2 receptor directs specificity of coupling to G-proteins. J Biol Chem 1995;270:7354–7358.
117 Asa SL, Kelly MA, Grandy DK, Low ML: Pituitary lactotroph adenomas develop after prolonged lactotroph hyperplasia in dopamine D2 receptor-deficient mice. Endocrinology 1999;140:5348–5355.
118 Sarkar DK, Chaturvedi K, Oomizu S, Boyadjieva N, Chen CP: Dopamine and TGF-β1 interact to inhibit the growth of pituitary lactotropes. Endocrinology 2005;146:4179–4188.
119 Nagineni CN, Samuel W, Nagineni S, Pardhasaradhi K, Wiggert B, Detrick B, Hooks JJ: Transforming growth factor-beta induces expression of vascular endothelial growth factor in human retinal pigment epithelial cells: involvement of mitogen-activated protein kinases. J Cell Physiol 2003;197:453–462.
120 Raymond V, Beaulieu M, Labrie F: Potent antidopaminergic activity of estradiol at the pituitary level on prolactin release. Science 1978;200:1173–1175.

Dipak K. Sarkar
Endocrine Program, Biomedical Division of the
Center of Alcohol Studies and Department of Animal Sciences
Rutgers, The State University of New Jersey
84 Lipman Drive
New Brunswick, NJ 08901–8525 (USA)
Tel. +1 732 932 1529, Fax +1 732 932 4134, E-Mail sarkar@AESOP.rutgers.edu

Arzt E, Bronstein M, Guitelman M (eds): Pituitary Today: Molecular, Physiological and Clinical Aspects. Front Horm Res. Basel, Karger, 2006, vol 35, pp 50–63

Dopaminergic D2 Receptor Knockout Mouse: An Animal Model of Prolactinoma

Carolina Cristina[a], *Isabel García-Tornadú*[a], *Graciela Díaz-Torga*[a], *Marcelo Rubinstein*[b], *Malcolm J. Low*[c], *Damasia Becú-Villalobos*[a]

[a]Instituto de Biología y Medicina Experimental-CONICET, [b]Instituto de Investigaciones en Ingeniería Genética y Biología Molecular, CONICET and University of Buenos Aires, Buenos Aires, Argentina; [c]The Vollum Institute, Oregon Health & Science University, Portland, Oreg., USA

Abstract

Dopamine receptor type 2 (D2R) knockout mice (KO) have chronic hyperprolactinemia, pituitary hyperplasia, and a moderate decrease in MSH content. They are also growth retarded evidencing an alteration in the GH-IGF-I axis. In D2R KO, lactotropes do not show dense secretory granules but degranulated cells and fewer somatotropes, gonadotropes and thyrotropes. Prolactin levels are always higher in female than in male knockouts, and in accordance, pituitary hyperplasia is observed at 8 months only in females. After 16 months of age, highly vascularized adenomas develop, especially in females. Prominent vascular channels in the hyperplastic and adenomatous pituitaries, as well as extravasated red blood cells not contained in capillaries is also a common finding. Prolactin is not the factor that enhances the hyperplastic phenotype in females while estrogen is a permissive factor. VEGF-A expression is increased in pituitaries from D2R KO. VEGF-A is expressed in follicle stellate cells. Because D2R receptors are found in lactotropes and not in follicle stellate cells, it may be inferred that a paracrine-derived factor from lactotropes is acting on follicle stellate cells to increase VEGF-A expression. VEGF-A does not induce pituitary cell proliferation, even though it enhances prolactin secretion. But it may act on adjacent endothelial cells and participate in the angiogenic process that increases the availability of different growth factors and mitogens. The D2R knockout mouse represents a unique animal model to study dopamine-resistant prolactinomas, and VEGF-A may be an alternative therapeutic target in this pathology.

Animal Models of Prolactinomas

Although mouse models may differ from their human counterparts, their study provides important insights into human pituitary tumor pathogenesis. There are several rat and mice models to study the genesis and regulation of prolactinomas, for example:

- The traditional estrogen-induced rat pituitary tumor.
- Transgenic mice with overexpression of a truncated fibroblast growth factor receptor 4 (FGFR4) isoform (driven by the prolactin promoter) [1].
- Targeted overexpression of transforming growth factor-α (TGF-α) under the control of the prolactin promoter [2].
- Transgenic nerve growth factor (NGF) overexpression under the control of the prolactin promoter [3].
- Overexpression of the high mobility group A (HMGA2) gene [4].
- Overexpression of galanin (driven by the GH or the prolactin promoter) [5, 6].
- Targeted disruption of TGFR type II (in heterozygotes) [7].
- The prolactin receptor knockout mouse [8].
- The dopamine (DA) receptor type 2 (D2R) knockout mouse [9, 10].

Each model presents variable characteristics: for example, the first three develop adenomas, while in NGF overexpression there is lactotrope hyperplasia with no adenoma formation. The overexpression of the truncated FGFR4 isoform is linked to diminished cell adhesion.

Dopaminergic Receptor Type 2 Knockout Mouse

We will focus on our results found in the D2R knockout mouse. DA is the most abundant catecholamine in the brain. The involvement and importance of DA as a neurotransmitter and neuromodulator in the regulation of different physiological functions in the central nervous system is well known. Deregulation of the dopaminergic system has been linked with Parkinson disease, Tourette syndrome, schizophrenia, attention-deficit hyperactive disorder and generation of pituitary tumors.

DA exerts its action by binding to specific membrane receptors, which belong to the family of seven transmembrane domain G-protein-coupled receptors. Five distinct DA receptors have been isolated, characterized and subdivided into two subfamilies, D1- and D2-like, on the basis of their biochemical and pharmacological properties. The D1-like subfamily comprises D1- and D5-R, while the D2-like includes the D2-, D3- and D4-Rs. In brain tissues the D2R is expressed predominantly in the caudate putamen, olfactory tubercle and nucleus accumbens. It is also expressed in the substantia nigra pars compacta and in the ventral tegmental area. These are the anatomical regions that give rise to long dopaminergic fibers (A10 y A9), indicating that the D2Rs have a presynaptic

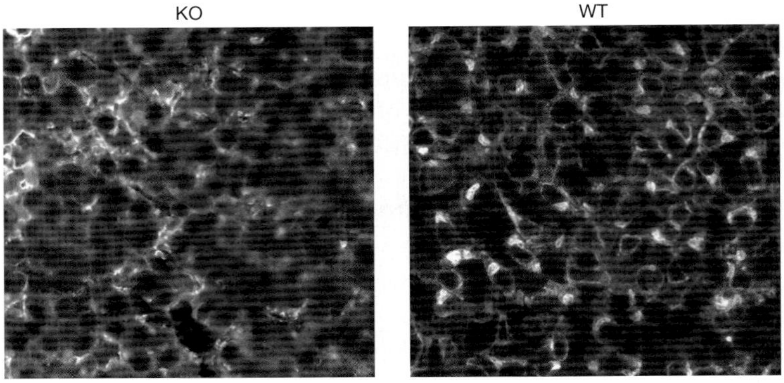

Fig. 1. Confocal microscopy of pituitaries from a D2R female knockout (left) and a wildtype mouse (right), using anti-PRL coupled to FITC.

location. Outside the brain D2R is localized in the retina, kidney, vascular system and in the pituitary gland, both in melanotropes and lactotropes.

The D2R gene is composed of eight exons, seven of which are coding. It encodes two splice variants, a short and a long D2R. The signal transduction pathways activated by D2Rs are numerous. They are coupled to pertussis toxin-sensitive Gi/Go proteins, and the best-described effects mediated by DA are the inhibition of the cAMP pathway and modulation of Ca^{2+} signaling.

Drs. Malcolm Low and Marcelo Rubinstein generated a D2R knockout mouse by targeted mutagenesis [10]. The study of locomotor behavior of D2R knockout mice revealed a motor impairment in mutant mice, even though mice did not show a compelling parkinsonian locomotor phenotype. Results predicted that locomotor activity is a polygenic trait that varies widely among inbred strains of mice [11]. On the other hand, this knockout mouse model yielded important results on the participation of the D2R in pituitary function. It has been well settled that the D2R is the principal receptor involved in prolactin inhibition at the pituitary level, and in MSH regulation at the intermediate pituitary.

Therefore, as expected D2R knockout mice had chronic hyperprolactinemia, pituitary hyperplasia, and a moderate decrease in MSH content [10]. They were also growth retarded evidencing an alteration in the GH-IGF-I axis [12].

In D2R knockout mice, lactotropes did not show the dense secretory granules characteristically found in wild types but degranulated cells, indicating constitutive secretion (fig. 1). Furthermore, we observed a decrease in somatotropes, gonadotropes and thyrotropes (fig. 2).

In knockout animals prolactin levels increased on the third month of life, and a chronic hyperprolactinemic state ensued throughout life [12]. Prolactin levels were always higher in female than in male knockout mice, and in accordance,

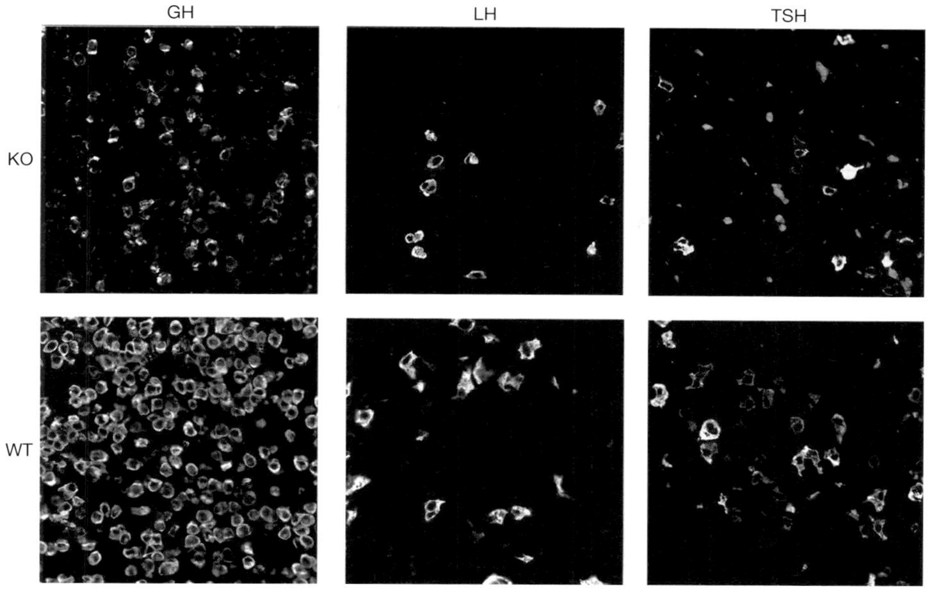

Fig. 2. Confocal microscopy of pituitaries from a D2R female knockout (upper panels: KO) and a wild-type mouse (lower panels: WT), using anti-GH, LH or TSH coupled to FITC.

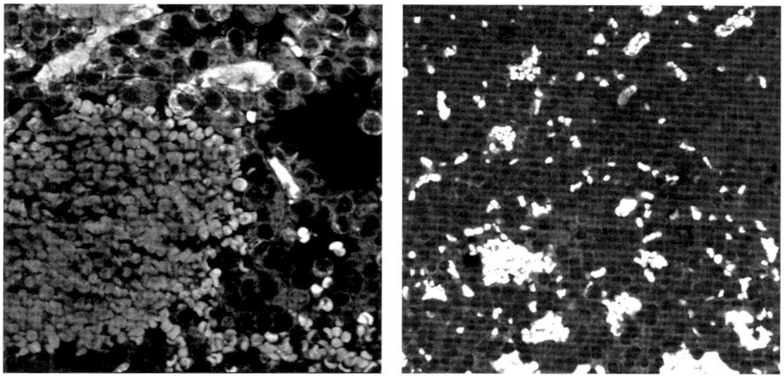

Fig. 3. Images of areas of peliosis in pituitaries from D2R knockout mice.

pituitary hyperplasia was observed at 8 months of age only in females [10]. After 16 months of age highly vascularized adenomas developed, especially in females [9]. Prominent vascular channels in the hyperplastic and adenomatous knockout pituitaries, as well as extravasated red blood cells not contained in capillaries or peliosis were also common findings (fig. 3).

Sex-Dependent Pituitary Hyperplasia in D2R Knockout Mice

Marked sexual differences in pituitary enlargement were important. As prolactin was chronically higher in females, it was possible that prolactin could perpetuate the hyperplastic phenotype by a positive feedback loop. To this regard, prolactin has been documented to be a proliferative growth factor in epithelial cells, kidney, the immune system and some tumors. A prolactin receptor (PRLR) knockout mouse was generated, and crossed with the D2R knockout. If prolactin was the proliferative factor acting at the pituitary, in the double knockout mouse hyperplasia would be reduced. But, contrary to the hypothesis PRLR knockout mice had enlarged pituitaries, and when combined with the D2R knockout pituitary hyperplasia was not decreased, and in males it was increased [8]. This clearly indicated that prolactin was not the factor involved in the sexual differences found in pituitary enlargement, and that at the pituitary level the prolactin receptor had an antiproliferative role.

A second candidate was estrogen, and indeed when female mice were ovariectomized, pituitary hyperplasia decreased. Nevertheless, estrogen replacement could not reproduce pituitary enlargement, indicating that estrogen was a permissive factor, but other factors were involved [13]. Furthermore, estrogen levels were not very different between male and female knockout mice [10].

We started to search for a factor which would participate in the generation of the hyperplasia in knockout females. Vascular endothelial growth factor-A (VEGF-A) seemed to be an interesting candidate as some reports have documented that it is regulated by DA and by estrogen [14–17], and besides, it is a potent angiogenic factor.

Neovascularization is essential for pituitary tumor formation, and is the result of a delicate balance between factors that stimulate or inhibit endothelial proliferation. A number of cytokines and growth factors have been demonstrated to modulate angiogenesis with a paracrine mode of action. Among these factors VEGF is one of the most potent angiogenesis inducers.

VEGF-A

VEGF-A is the founding member of a family of closely related cytokines that exert critical functions in vasculogenesis and in both pathologic and physiologic angiogenesis and lymphangiogenesis [18]. The VEGF-A gene is located on the short arm of chromosome 6 and is differentially spliced to yield several different isoforms, the three most prominent of which encode polypeptides of

189, 165 and 121 amino acids in human cells. The protein has a hydrophobic leader sequence, typical of secreted proteins. It was discovered in the late 1970s as a tumor-secreted protein that potently increased microvascular permeability to plasma proteins. VEGF-A is a potent mitogen for micro- and macrovascular endothelial cells derived from arteries, veins and lymphatics but not for other cell types [18]. Thus, in addition to rendering microvessels hyperpermeable, VEGF-A stimulates endothelial cells to migrate and divide.

We can summarize its unique properties [19]:

(1) It is essential for normal developmental vasculogenesis and angiogenesis, as both null (VEGF-A−/−) and heterozygote (VEGF-A±) animals are embryonic lethals.

(2) It increases vascular permeability to plasma and plasma proteins, a characteristic property of the tumor microvasculature and a critical early step in tumor stroma generation.

(3) It is selective for vascular endothelium because its major tyrosine kinase receptors are selectively (though not exclusively) expressed on vascular endothelium.

(4) It is overexpressed in a variety of human cancer cells. Enhanced VEGF-A expression has been associated with several human vascular tumors including brain, colon, gastrointestinal tract, ovary, breast and others.

(5) It has a potential for evaluating prognosis in individual patients and as a therapeutic target.

Recently, it was found, quite unexpectedly, that DA and other related catecholamine neurotransmitters that interact with the D2R selectively inhibit VEGF-A-induced angiogenesis and inhibit the growth of tumors that express VEGF-A. At nontoxic doses, DA inhibited all VEGF-A activities tested including stimulation of microvascular permeability and endothelial cell proliferation and migration [14, 20]. Besides, in two outbred lines of Wistar rats, which present high and low dopaminergic reactivity, respectively, VEGF expression was reduced in the first group, which was more resistant to tumor implantation, and developed significantly fewer lung metastases [20]. These data point to a relation of VEGF-A and DA receptors.

Interesting to our experimental model was also the fact that the VEGF-A gene promoter has AP-1 and AP-2 sites. This means that forskolin which activates adenylate cyclase and increase cAMP induce VEGF mRNA expression [19]. In the pituitary of the D2R knockout mice, the lack of action of DA on its receptor prevents physiological adenylate cyclase inhibition, probably leading to increased cAMP levels.

Therefore, because VEGF is thought to be the most important angiogenic cytokine in cancer and other types of pathological angiogenesis and because it has been related to the D2R in endothelial cells [14, 15] we investigated VEGF expression, localization, and function in pre-adenomatous pituitary tumors of D2R knockout female mice.

VEGF-A in Pituitaries from D2R Knockout Female Mice

We found that VEGF expression was increased in pituitaries from D2R knockout female mice when compared to age-matched, wild-type female mice [21]. VEGF production had been demonstrated to be stimulated by estrogen in rat pituitaries [16] and the somatolactotrope cell line GH3 [22]. Nevertheless, estrogen levels are not increased in D2R knockout female mice, indicating that increased pituitary VEGF expression is mainly dependent on the lack of dopaminergic control. In experiments with wild-type female mice we found that prolonged treatment with the D2R antagonist, haloperidol, enhanced pituitary VEGF protein content and prolactin release [21], and that there was a significant correlation between pituitary VEGF levels and serum prolactin after haloperidol treatment (fig. 4). This further supports that DA acting at the D2R inhibits pituitary VEGF expression. VEGF was also increased in cultured pituitary cells from knockout mice, and in the conditioned media from these cells.

Pituitary VEGF-A Localization

In the normal human pituitary, VEGF-A has been localized mainly in ACTH, GH and follicle stellate cells, with lower levels detected in other cell types [23]. In rats, VEGF-A has been described in follicle stellate cells [24], in a part of the total TSH cells [25], as well as in the lactosomatotrope GH3 pituitary tumor cell line [22]. Interestingly, we found that the main source of VEGF-A in the hyperplastic pituitary were follicle stellate cells and not lactotropes [21].

Follicle stellate cells represent 5–10% of pituitary cells and are an important component of paracrine communication within the pituitary. The expression of gap-junction channels in follicle stellate cells supports the hypothesis that they operate as a network of functionally interconnected cells [26]. They were originally described as non hormone secreting cells, and they are believed to function as (1) support; (2) fagocytes; (3) trophic cells; (4) stem cells; (5) transport cells, and (6) paracrine modulators. They are detected by their content of the glial protein S100, they form follicles, are star shaped and have long processes in between the secretory cells of the pituitary. They also contain FGF-2, follistatin and interleukin-6.

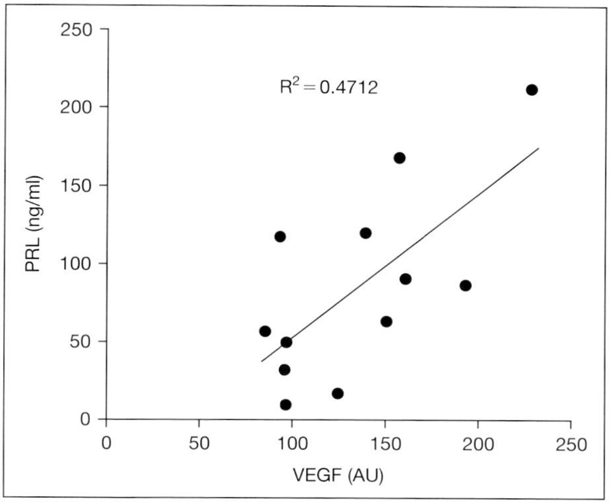

Fig. 4. Correlation between serum prolactin levels (measured by RIA), and pituitary VEGF-A content (Western blot, normalized to actin content, AU = arbitrary units), after i.p. administration of haloperidol decanote (5 mg/kg s.c. for 3 weeks, one injection per week, or 1.2 mg/kg i.p. for 7 days, one injection per day).

Because D2R receptors have been described in lactotropes and not in follicle stellate cells, it may be inferred that a paracrine-derived factor from lactotropes is acting on follicle stellate cells to increase VEGF-A expression. To this regard, it has been described that agents that increase cAMP levels increase VEGF-A in a follicle stellate cell line [27]. As mentioned, in D2R knockout lactotropes, DA-mediated inhibition of adenylate cyclase is chronically lacking, so we presume that increased production of cAMP in this cell type may modulate VEGF-A synthesis in neighboring follicle stellate cells.

VEGF-A and Peliosis

Another interesting association found was that of VEGF-A and peliosis (extravasated erythrocytes not contained in capillaries). As mentioned, in hyperplastic and in adenomatous pituitaries of female D2R knockout mice we described the occurrence of peliosis. Peliosis has been found in different tumors that secrete VEGF-A. This may be linked to the permeabilizing function of this growth factor, and to the increased fenestration induced in blood vessels stimulated by VEGF-A overexpression. Peliosis occurrence has been related to high

VEGF-A expression in hepatocarcinogenesis, spleen damage and in a lethal hepatic syndrome in mice [28–30].

VEGF-A Action at the Pituitary Level

We also determined whether VEGF-A had any action on pituitary cell proliferation or prolactin release. VEGF-A mediates its mitogenic and vasopermeabilizing effects principally through two tyrosine kinase receptors, VEGF-R1 (or Flt 1) and VEGF-R2 (KDR, or Flk-1). Both VEGF-R1 and VEGF-R2 have seven immunoglobulin (Ig)-like domains in the extracellular domain. VEGF receptor-2 has the highest binding affinity for VEGF-A [19].

Expression of these two VEGF receptors exclusively on endothelial cells indicates that this factor should have no direct influence on endocrine cells. VEGF-A might act on the intrapituitary endothelium, maintaining vascular integrity, stimulating vascular permeability and endothelial cell proliferation. Nevertheless, there is one report of VEGF-R2 expression in pituitary endocrine cells [31].

We found that VEGF-A did not induce pituitary cellular proliferation; this result is consistent with several reports that claim that VEGF-A is a potent mitogen for vascular endothelial cells but that it is devoid of consistent and appreciable mitogenic activity for other cell types [32]. In fact, the denomination of VEGF was proposed to emphasize such narrow target cell specificity.

On the other hand, a prolactin-releasing effect could be evidenced [21]. The prolactin-releasing effect of VEGF-A could be related to VEGF-R described in the pituitary cells. This effect was evidenced only under an estrogenic environment. To this respect, it has been conclusively described that estradiol modifies lactotrope sensitivity to physiological stimulators and inhibitors of prolactin secretion.

VEGF-A and Angiogenesis

Therefore, increased pituitary VEGF-A expression may not be important for cellular proliferation of endocrine cells per se, even though it may enhance the prolactin secretory capacity of the gland. On the other hand, increased VEGF-A may act in adjacent endothelial cells and participate in the angiogenic process that increases the availability of different growth factors and mitogens.

To support this idea, we found that conditioned medium from the hyperplastic pituitaries (D2R−/−) was able to induce proliferative changes in human umbilical vein endothelial cells (HUVECs) [21]. The proliferating effect was in part evoked by secreted VEGF-A, as shown by immunoneutralization experiments. This probably indicates that pituitary-secreted VEGF-A accumulates in the target endothelial cells in which it may act in a paracrine manner enhancing vessel proliferation.

Role of Cell Adhesion in Pituitary Tumorigenesis

Cell adhesion is an important determinant of organized growth and the maintenance of architectural integrity within the pituitary. Reduced adhesiveness between cells and, between cells and the extracellular matrix (ECM) is a hallmark of neoplastic growth. The ECM is a three-dimensional network of proteins, glycosaminoglycans and other macromolecules which has a structural support function as well as a role in cell adhesion, migration, proliferation, differentiation, and survival. The ECM conveys signals through membrane receptors called integrins and plays an important role in pituitary physiology and tumorigenesis.

To this regard, the pituitary contains some components within the basement membrane, such as laminin and FGF-2, which decrease in the tumorigenic process in the D2R knockout mouse [33, 34]. These reductions could reflect a more general reduction of the basement membrane by action of matrix metalloproteinases. Matrix metalloproteinase activity is very high in all types of human pituitary adenomas. These metalloproteinases secreted by pituitary cells could release growth factors from the ECM that, in turn, may control pituitary cell proliferation and hormone secretion. Therefore, we must bear in mind that FGF2 and laminin might be released in the process of ECM remodeling and participate in proliferation and angiogenesis in an early step of tumor formation. Remodeling of the existing ECM and diminished cell adhesiveness has been linked to pituitary tumorigenesis in transgenic mice with overexpression of a truncated FGFR4 isoform [35].

D2R Knockout Mouse as an Experimental Model for Dopamine-Resistant Prolactinomas

We have described that pituitary VEGF-A and not FGF2 expression is increased in female mice lacking DA D2Rs. Even though VEGF-A does not promote pituitary cellular proliferation in vitro, as it does in endothelial cells, it may be critical for effective tumor angiogenesis, which is important for pituitary hyperplasia, and, furthermore, it may participate in increased prolactin secretion.

Prolactin adenomas are common monoclonal neoplasms accounting for approximately 30% of intracranial tumors. They are usually benign, and can be effectively treated with dopaminergic agents. But 15% of these may become resistant to classical pharmacological therapy, are invasive and aggressive, and require extirpation. An alternative target, such as VEGF-A, would be desired in these circumstances.

It has been shown that tumor characteristics and environment promote VEGF-A expression. For example, oncogene expression (*kRAS*, *HRAS* and *HER2*), EGF, TGF-β, or keratinocyte growth factor and hypoxia have been related to VEGF-A regulation. The description of dopaminergic control of VEGF-A expression in the pituitary may be important in the clinical action of dopaminergic agents. Furthermore, we believe that VEGF-A and its receptor may become important therapeutic tools in DA-resistant prolactinomas.

To this regard, in recent years, antiangiogenesis has been publicized as a novel alternative or supplement to conventional cancer therapy, and a variety of regimens that prevent tumor angiogenesis and/or that attack tumor blood vessels have met with remarkable success in treating mouse cancers.

Overexpression of VEGF-A by tumor cells can be targeted by:

- antibodies against VEGF (Bevacizumab);
- antibodies against VEGF receptors;
- soluble VEGF receptors (VEGF-TRAP) that bind circulating VEGF;
- catalytic RNA molecules (ribozymes), which cleave VEGF receptor mRNA;
- orally available molecules that selectively block or prevent activation of VEGF-A's receptor tyrosine kinases.

Despite the spectacular successes reported in the treatment of mouse tumors, the first clinical trials were discouragingly negative. This could be related to the fact that most of the patients treated in the beginning had advanced disease and had already failed conventional treatments. Also, antiangiogenesis therapy differs fundamentally from chemotherapy, and optimal implementation was needed.

Several agents targeting the VEGF ligand are now being developed in different clinical trials around the world to treat colon, rectal, breast, lung and other cancers [36]. Bevacizumab (Avastin™), an anti-VEGF monoclonal antibody that inhibits formation of neovasculature and tumor growth in many human cancer cell lines, has a proven survival benefit in metastatic colon rectal cancer and has now been approved by the FDA in combination with intravenous 5-FU-based chemotherapy as a treatment for patients with first-line metastatic cancer of the colon or rectum [37].

Angiogenesis and Pituitary Hyperplasia

It is important to consider that normal pituitaries are usually highly vascularized; therefore, the changes that occur during tumor development may be somewhat different from some other tissues which require active angiogenesis. To this regard, slow angiogenesis in prolactinomas and even lower vascular densities have been described in human adenomas compared to normal pituitary tissue [38, 39]. Only carcinomas exhibit a clear-cut increase in microvascular density, and there is no information about the angiogenic process in DA-resistant prolactinomas.

Conclusions

In conclusion, we describe that mice lacking DA D2Rs present an enhanced pituitary VEGF-A expression in correlation with pituitary hyperplasia and peliosis of the gland. FGF2, on the other hand, is decreased and such a decrease may be related to the low angiogenesis in prolactinomas and may have a role in the slow pace of pituitary tumor growth, the rarity of metastases [38], and the benign nature of the tumors.

Lack of the D2R likely encourages or permits the development of tumors by increasing the population of proliferating cells that are susceptible to oncogenic factors, or mutation. Increased VEGF-A expression may participate in enhancing permeability and maintaining angiogenesis. This would be in accordance with the multistep theory of carcinogenesis, which reconciles the two proposed theories of pituitary tumorigenesis (hormonal or clonal expansion). It is likely that the majority of pituitary adenomas develop from transformed cells that are, to some extent, dependent on hormonal and/or growth factor stimulation for tumor progression. Proliferating cells would be at increased risk of genetic alteration during mitosis and manifestation of genetic alterations would be precipitated by the growth stimulus.

References

1 Ezzat S, Zheng L, Zhu XF, Wu GE, Asa SL: Targeted expression of a human pituitary tumor-derived isoform of FGF receptor-4 recapitulates pituitary tumorigenesis. J Clin Invest 2002:109: 69–78.

2 McAndrew J, Paterson AJ, Asa SL, McCarthy KJ, Kudlow JE: Targeting of transforming growth factor-alpha expression to pituitary lactotrophs in transgenic mice results in selective lactotroph proliferation and adenomas. Endocrinology 1995:136:4479–4488.

3 Borrelli E, Sawchenko PE, Evans RM: Pituitary hyperplasia induced by ectopic expression of nerve growth factor. Proc Natl Acad Sci USA 1992:89:2764–2768.

4 Fedele M, Battista S, Kenyon L, Baldassarre G, Fidanza V, Klein-Szanto AJ, Parlow AF, Visone R, Pierantoni GM, Outwater E, Santoro M, Croce CM, Fusco A: Overexpression of the HMGA2 gene in transgenic mice leads to the onset of pituitary adenomas. Oncogene 2002:21:3190–3198.

5 Perumal P, Vrontakis ME: Transgenic mice over-expressing galanin exhibit pituitary adenomas and increased secretion of galanin, prolactin and growth hormone. J Endocrinol 2003:179: 145–154.

6 Cai A, Hayes JD, Patel N, Hyde JF: Targeted overexpression of galanin in lactotrophs of transgenic mice induces hyperprolactinemia and pituitary hyperplasia. Endocrinology 1999:140:4955–4964.
7 Ikeda H, Yoshimoto T, Shida N, Miyoshi I, Nakayama K, Nakayama K, Oshima M, Taketo MM: Morphologic and molecular analysis of estrogen-induced pituitary tumorigenesis in targeted disruption of transforming growth factor-beta receptor type II and/or p27 mice. Endocrine 2001:16:55–65.
8 Schuff KG, Hentges ST, Kelly MA, Binart N, Kelly PA, Iuvone PM, Asa SL, Low MJ: Lack of prolactin receptor signaling in mice results in lactotroph proliferation and prolactinomas by dopamine-dependent and -independent mechanisms. J Clin Invest 2002:110:973–981.
9 Asa SL, Kelly MA, Grandy DK, Low MJ: Pituitary lactotroph adenomas develop after prolonged lactotroph hyperplasia in dopamine D2 receptor-deficient mice. Endocrinology 1999:140: 5348–5355.
10 Kelly MA, Rubinstein M, Asa SL, Zhang G, Saez C, Bunzow JR, Allen RG, Hnasko R, Ben-Jonathan N, Grandy DK, Low MJ: Pituitary lactotroph hyperplasia and chronic hyperprolactinemia in dopamine D2 receptor-deficient mice. Neuron 1997:19:103–113.
11 Kelly MA, Rubinstein M, Phillips TJ, Lessovn CN, Burkhart-Kasch S, Zhang G, Bunzow JR, Fang Y, Gerhardt GA, Grandy DK, Low MJ: Locomotor activity in D2 dopamine receptor-deficient mice is determined by gene dosage, genetic background, and developmental adaptations. J Neurosci 1998:18:3470–3479.
12 Diaz-Torga G, Feierstein C, Libertun C, Gelman D, Kelly MA, Low MJ, Rubinstein M, Becu-Villalobos D: Disruption of the D2 dopamine receptor alters GH and IGF-I secretion and causes dwarfism in male mice. Endocrinology 2002:143:1270–1279.
13 Hentges ST, Low MJ: Ovarian dependence for pituitary tumorigenesis in D2 dopamine receptor-deficient mice. Endocrinology 2002:143:4536–4543.
14 Basu S, Nagy JA, Pal S, Vasile E, Eckelhoefer IA, Bliss VS, Manseau EJ, Dasgupta PS, Dvorak HF, Mukhopadhyay D: The neurotransmitter dopamine inhibits angiogenesis induced by vascular permeability factor/vascular endothelial growth factor. Nat Med 2001:7:569–574.
15 Chakroborty D, Sarkar C, Mitra RB, Banerjee S, Dasgupta PS, Basu S: Depleted dopamine in gastric cancer tissues: dopamine treatment retards growth of gastric cancer by inhibiting angiogenesis. Clin Cancer Res 2004:10:4349–4356.
16 Banerjee SK, Sarkar DK, Weston AP, De A, Campbell DR: Over expression of vascular endothelial growth factor and its receptor during the development of estrogen-induced rat pituitary tumors may mediate estrogen-initiated tumor angiogenesis. Carcinogenesis 1997:18:1155–1161.
17 Onofri C, Carbia NA, Schaaf L, Feirer M, Lohrer P, Stummer W, Berner S, Chervin A, Goldberg V, Stalla GK, Renner U, Arzt E: Estradiol stimulates vascular endothelial growth factor and interleukin-6 in human lactotroph and lactosomatotroph pituitary adenomas. Exp Clin Endocrinol Diabetes 2004:112:18–23.
18 Ferrara N, Gerber H, LeCouter J: The biology of VEGF and its receptors. Nat Med 2003:9: 669–676.
19 Ferrara N, Davis-Smyth T: The biology of vascular endothelial growth factor. Endocr Rev 1997:18: 4–25.
20 Teunis MA, Kavelaars A, Voest E, Bakker JM, Ellenbroek BA, Cools AR, Heijnen CJ: Reduced tumor growth, experimental metastasis formation, and angiogenesis in rats with a hyperreactive dopaminergic system. FASEB J 2002:16:1465–1467.
21 Cristina C, Diaz-Torga G, Baldi A, Gongora A, Rubinstein M, Low MJ, Becu-Villalobos D: Increased pituitary vascular endothelial growth factor-A in dopaminergic D2 receptor knockout female mice. Endocrinology 2005:146:2952–2962.
22 Ochoa AL, Mitchner NA, Paynter CD, Morris RE, Ben Jonathan N: Vascular endothelial growth factor in the rat pituitary: differential distribution and regulation by estrogen. J Endocrinol 2000:165:483–492.
23 Vidal S, Kovacs K, Cohen SM, Stefaneanu L, Lloyd RV, Scheithauer BW: Localization of vascular endothelial growth factor in non tumorous human pituitaries. Endocr Pathol 1999:10: 109–122.

24 Gospodarowicz D, Abraham JA, Schilling J: Isolation and characterization of a vascular endothelial mitogen produced by pituitary-derived folliculostellate cells. Proc Natl Acad Sci USA 1989:86:7311–7315.
25 Fan L, Iseki S: Immunohistochemical localization of vascular endothelial growth factor in the endocrine glands of the rat. Arch Histol Cytol 1998:61:17–28.
26 Stojilkovic SS: A novel view of the function of pituitary folliculo-stellate cell network. Trends Endocrinol Metab 2001:12:378–380.
27 Gloddek J, Pagotto U, Paez PM, Arzt E, Stalla GK, Renner U: Pituitary adenylate cyclase-activating polypeptide, interleukin-6 and glucocorticoids regulate the release of vascular endothelial growth factor in pituitary folliculostellate cells. J Endocrinol 1999:160:483–490.
28 Joseph F, Younis N, Haydon G, Adams DH, Wynne S, Gillet MB, Maurice YM, Lipton ME, Berstock D, Jones IR: Peliosis of the spleen with massive recurrent haemorrhagic ascites, despite splenectomy, and associated with elevated levels of vascular endothelial growth factor. Eur J Gastroenterol Hepatol 2004:16:1401–1406.
29 Park YN, Kim YB, Yang KM, Park C: Increased expression of vascular endothelial growth factor and angiogenesis in the early stage of multistep hepatocarcinogenesis. Arch Pathol Lab Med 2000:124:1061–1065.
30 Wong AK, Alfert M, Castrillon DH, Shen Q, Holash J, Yancopoulos GD, Chin L: Excessive tumor-elaborated VEGF and its neutralization define a lethal paraneoplastic syndrome. Proc Natl Acad Sci USA 2001:98:7481–7486.
31 Vidal S, Lloyd RV, Moya L, Scheithauer BW, Kovacs K: Expression and distribution of vascular endothelial growth factor receptor Flk-1 in the rat pituitary. J Histochem Cytochem 2002:50: 533–540.
32 Yu R, Ren S-G, Horwitz GA, Wang Z, Melmed S: Pituitary tumor transforming gene (PTTG) regulates placental JEG-3 cell division and survival: evidence from live cell imaging. Mol Endocrinol 2000:14:1137–1146.
33 Kuchenbauer F, Theodoropoulou M, Hopfner U, Stalla J, Renner U, Tonn JC, Low MJ, Arzt E, Stalla GK, Paez-Pereda M: Laminin inhibits lactotroph proliferation and is reduced in early prolactinoma development. Mol Cell Endocrinol 2003:207:13–20.
34 Cristina C, Achaval-Zaia R, Baldi A, Low MJ, Rubinstein M, Diaz-Torga G, Becu-Villalobos D: Factores de crecimiento en la hiperplasia hipofisaria de ratones hembra deficientes en el receptor dopaminérgico tipo 2. Medicina 2000:62:469.
35 Ezzat S, Zheng L, Asa SL: Pituitary tumor-derived fibroblast growth factor receptor 4 isoform disrupts neural cell-adhesion molecule/N-cadherin signaling to diminish cell adhesiveness: a mechanism underlying pituitary neoplasia. Mol Endocrinol 2004:18:2543–2552.
36 Bergsland EK: Update on clinical trials targeting vascular endothelial growth factor in cancer. Am J Health Syst Pharm 2004:61:S12–S20.
37 Hurwitz H, Fehrenbacher L, Novotny W, Cartwright T, Hainsworth J, Heim W, Berlin J, Baron A, Griffing S, Holmgren E, Ferrara N, Fyfe G, Rogers B, Ross R, Kabbinavar F: Bevacizumab plus irinotecan, fluorouracil, and leucovorin for metastatic colorectal cancer. N Engl J Med 2004:350:2335–2342.
38 Yamada S, Takada K: Angiogenesis in pituitary adenomas. Microsc Res Tech 2003:60:236–243.
39 Lloyd RV, Vidal S, Horvath E, Kovacs K, Scheithauer B: Angiogenesis in normal and neoplastic pituitary tissues. Microsc Res Tech 2003:60:244–250.

Damasia Becú-Villalobos
Instituto de Biología y Medicina Experimental, CONICET
Vuelta de Obligado 2490
1428 Buenos Aires (Argentina)
Tel. +54 11 47832869, Fax +54 11 47862564, E-Mail dbecu@dna.uba.ar

Arzt E, Bronstein M, Guitelman M (eds): Pituitary Today: Molecular, Physiological and Clinical Aspects. Front Horm Res. Basel, Karger, 2006, vol 35, pp 64–73

Vasoinhibins: A Family of N-Terminal Prolactin Fragments that Inhibit Angiogenesis and Vascular Function

Carmen Clapp, Carmen González, Yazmín Macotela, Jorge Aranda, José C. Rivera, Celina García, Jessica Guzmán, Miriam Zamorano, Claudia Vega, Cecilia Martín, Michael C. Jeziorski, Gonzalo Martínez de la Escalera

Instituto de Neurobiología, Universidad Nacional Autónoma de México, Campus UNAM-Juriquilla, Querétaro, Qro, México

Abstract

Antiangiogenic molecules derived from prolactin (PRL) are not a single entity, but rather a family of peptides with different molecular masses, all containing the N-terminal region of PRL. Cleavage of PRL by cathepsin-D or by matrix metalloproteases generates N-terminal fragments that act on endothelial cells to suppress vasodilation and angiogenesis and promote vascular regression. N-terminal PRL fragments have been identified in cartilage and retina, where angiogenesis is highly restricted. In vivo experiments demonstrate that these PRL fragments exert a tonic and essential suppression of retinal blood vessel growth and dilation. Similar PRL fragments have been detected in the pituitary gland, a highly vascularized organ where the control of vascular growth may differ from that in tissues where angiogenesis is highly restricted. We have previously proposed the name vasoinhibins to describe the collection of N-terminal PRL fragments having blood vessel-blocking activity, and here we discuss their promise as factors to control vascular function in health and disease.

Angiogenesis, the formation of new blood vessels from the existing vasculature, is essential for tissue growth during development, but is almost nonexistent in the adult except in wound repair and female reproductive events. The quiescence of vasculature is dependent upon a local balance between pro- and antiangiogenic factors that, when disrupted, leads to pathological angiogenesis in diseases such as vasoproliferative retinopathies, rheumatoid arthritis, and the growth and spread of tumors. Therefore, the ability to inhibit angiogenesis offers a unique opportunity to regulate a wide variety of physiological and pathological states.

Among the many known antiangiogenic factors, at least half belong to a family of peptides that become active only after they are released by proteolytic cleavage of large precursor proteins [1]. These proteolytic fragments can interfere with various phases of the angiogenesis process by inhibiting endothelial cell migration, proliferation, differentiation, and survival, and include angiostatin, endostatin, cleaved-antithrombin, tumstatin, canstatin, vasostatin, restin, arrestin and N-terminal fragments of PRL [1, 2]. While these pluripotent factors may provide new therapeutic agents for combating vascular disease, knowledge of their basic biology and mechanism of action is a prerequisite for developing decisive targeting strategies.

Vasoinhibins

Mature PRL is a 23-kDa molecule of 199 (human) or 197 (rat) amino acids. The antiangiogenic properties of N-terminal PRL fragments were discovered nearly 15 years ago in association with the 16-kDa fragment comprising residues 1–145 that results from cleavage of rat prolactin by cathepsin-D [3, 4]. Subsequently, several N-terminal fragments of human PRL produced by recombinant technology were shown in various bioassays to display antiangiogenic properties similar to those reported for the 16 kDa rat protein [5–7]. More recently, 17 kDa fragments of both human and rat PRL generated by cleavage with matrix metalloproteases (MMP) were found to exert antiangiogenic effects [8].

Fragments generated by cleavage of PRL at different sites require the N-terminal end of the molecule to actively inhibit angiogenesis, since a 16-kDa C-terminal fragment of PRL (residues 54–199) produced by thrombin is devoid of antiangiogenic properties [9]. Known N-terminal PRL fragments with antiangiogenic activity range in length from 124 to 155 amino acids with calculated masses between 14.1 and 17.3 kDa (fig. 1), all lacking the fourth alpha-helix of mature PRL [10]. Because the initial study characterizing antiangiogenic effects was carried out with a 16-kDa N-terminal fragment of rat PRL [3], 16K-PRL was chosen as a single name, and subsequent studies continued to identify the active molecules as 16K-PRL, even though they involved PRL fragments of different masses [5–8]. However, it is clearly inaccurate to consider antiangiogenic PRL as a single, homogenous protein fragment. We have thus proposed the name vasoinhibins to define the collection of N-terminal fragments of PRL having inhibitory actions on vascular endothelial cells [8].

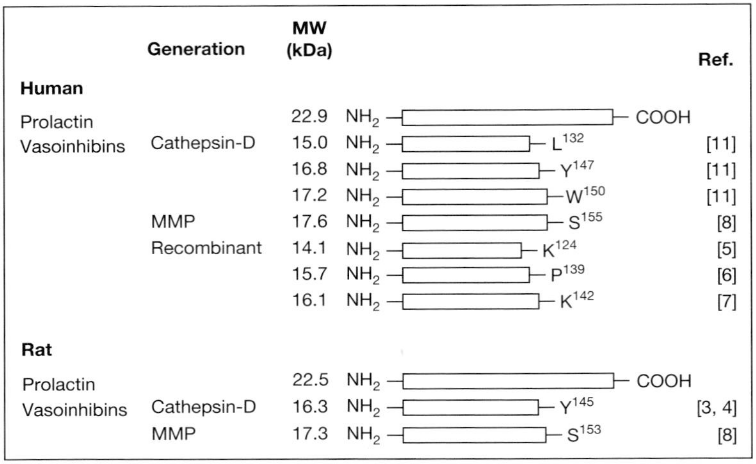

Fig. 1. Endogenous and recombinant fragments of human and rat prolactin known to inhibit angiogenesis. The parent molecules are included for comparison. Predicted molecular weights are calculated from amino acid sequences.

Generation of Vasoinhibins

Several proteases appear to mediate the generation of vasoinhibins, as is also the case for the other cryptic antiangiogenic factors [1]. Cathepsin-D was the first protease claimed to be responsible for cleaving PRL to a vasoinhibin of 16 kDa [4]. However, the relevance of this enzyme is debatable since human PRL, unlike rat PRL, is highly resistant to cleavage by cathepsin-D [9, 11]. Likewise, the neutral pH of physiological fluids contrasts with the acidic conditions normally required for the proteolytic activity of this enzyme. More recently, MMP secreted by chondrocytes and various purified MMP (MMP-1, -2, -3, -8, -9 and -13) that are active at neutral pH were shown to be equally effective at cleaving PRL from various species into vasoinhibins [8]. While little is known about the specific proteases responsible for generating vasoinhibins, the existence of the endogenous peptides [8, 12–15] suggests that such enzymes are active in vivo.

Endogenous Vasoinhibins

Evidence indicates that PRL fragments with apparent molecular masses of 14, 16 and/or 17 kDa occur naturally in the anterior and posterior pituitary gland [12, 14, 15], the hypothalamus [16], the retina [13], vascular endothelial cells [17], pulmonary fibroblasts [18], chondrocytes [8], and PRL-secreting adenomas [11]. These fragments correspond to the N-terminal part of the PRL molecule, as they are recognized in Western blots by vasoinhibin-directed polyclonal antibodies and by monoclonal antibodies against the N-terminal end of PRL but not by those directed against the C-terminus of the hormone. Likewise, immunoreactive 12-, 14- and/or 16-kDa PRL fragments have been detected in serum [14, 19], amniotic fluid [20], and subretinal fluid [21] from humans or rats. However, the N- or C-terminal nature of PRL fragments in biological fluids has not yet been carefully examined.

Vascular Properties of Vasoinhibins

Angiogenesis initiates with vasodilation and increased vessel permeability, allowing extravasation of plasma proteins that lay down a temporary scaffold for migrating endothelial cells. Subsequently, endothelial cells proliferate and assemble into tubes that become stabilized by interaction with periendothelial cells (pericytes and smooth muscle cells). Once formed, the new vessels become quiescent and survive for years, except when vascular regression is required. For example, endothelial cell apoptosis causes regression of blood vessels in order to remodel the vascular network during development and after birth in the retina and ovary [22, 23]. Also, vascular regression can occur during pathological angiogenesis, limiting the progression of angiogenesis-dependent diseases [22, 24].

Like most antiangiogenic factors, vasoinhibins interfere with various steps of the angiogenic process. They inhibit angiogenesis in vivo and in vitro, suppressing growth factor-induced vasodilation [25] and endothelial cell proliferation [5, 26], and stimulating the expression of the type 1 plasminogen activator inhibitor [27]. Furthermore, vasoinhibins stimulate endothelial cell apoptosis [28], causing the regression of blood vessels [21]. The molecular mechanisms underlying these inhibitory actions include blocking the MAPK pathway [29], inhibiting the activation of endothelial (eNOS) and inducible (iNOS) nitric oxide synthases [25, 30], and activating the caspase pathway via NFκB [7, 31]. However, as for many other anti-angiogenic proteolytic fragments, the receptor for vasoinhibins has not been identified [32], and the signaling molecules that mediate the antiangiogenic activity of vasoinhibins remain poorly characterized [12].

The fact that vasoinhibins inhibit nitric oxide (NO) production by endothelial cells [25, 30] suggests that, besides angiogenesis, other vascular events may be downregulated by these peptides. NO produced by eNOS activation is physiologically important for maintaining vascular homeostasis; it keeps the vessels dilated, protects the intima from adhesion of platelet aggregates and leukocytes, and prevents proliferation and migration of smooth muscle cells [33]. In this regard, vasoinhibins prevent vessel dilation [25], whereas the uncleaved PRL molecule stimulates adhesion of leukocytes to vascular endothelial cells [34].

In vivo Activity of Vasoinhibins

The facts that in vitro proteolytic cleavage produces vasoinhibins and that vasoinhibins occur naturally are consistent with a role of this family of peptides in angiogenesis. However, this role could not be confirmed unless endogenous vasoinhibins were demonstrated to be active in vivo. Targeted disruption of either PRL [35] or the PRL receptor [36] has not been particularly useful for dissecting the in vivo role of vasoinhibins. The PRL gene was mutated by a targeted insertion that truncated the protein at residue 117, leaving an intact N-terminal PRL fragment of 11 kDa [35] that may retain antiangiogenic properties and therefore no change of the vasoinhibin phenotype. Also, vasoinhibins appear to signal through a cell surface receptor in endothelial cells that is distinct from the PRL receptor [32], so targeted deletion of the PRL receptor was not expected to interfere with its activity.

We have used the eye as a model to investigate the in vivo role of endogenous vasoinhibins. In the eye, vessels are normally excluded from the cornea, the vitreous, the lens, and the outer half of the retina. Failure to exclude blood vessels can result in reduced visual acuity, opacification, and abnormal healing, whereas excessive retinal neovascularization underlies premature retinopathy, diabetic retinopathy, and age-related macular degeneration, the leading causes of blindness throughout the world [37]. Evidence indicates that vasoinhibins are present in ocular tissues. PRL is found in the aqueous humor of the human eye [21, 38], and both PRL and vasoinhibins are detected in rat retina [13]. Some ocular PRL may originate from systemic PRL entering the eye, as iodinated PRL injected intracardially can be incorporated into the retina, choroid, and ciliary body [39]. However, PRL and vasoinhibins can also be synthesized intraocularly. PRL mRNA is expressed in rat retina [13], where vascular endothelium is its likely source. Cultures of rat retinal capillary endothelial cells synthesize and secrete PRL [40], and blood vessels from retrolental fibrovascular membranes of patients with retinopathy of prematurity express PRL mRNA [21]. In addition, ocular fluids contain the cleaving enzymes able to generate vasoinhibins from PRL [21].

Within the eye, vasoinhibins block the stimulatory action of various inducers of angiogenesis. Gene transfer of vasoinhibins via an adenoviral vector significantly inhibits retinal angiogenesis in murine ischemia-induced retinopathy

[41], and local administration of vasoinhibins reduces the stimulation of corneal angiogenesis by basic fibroblast growth factor [42]. Moreover, the endogenous peptide suppresses angiogenesis in healthy ocular tissue. Intracorneal implantation of a pellet releasing antibodies able to inactivate endogenous vasoinhibin resulted in vessel growth into the normally avascular cornea [42]. Furthermore, intravitreal injection of antibodies against vasoinhibins stimulated vessel growth in the retina, and intraocular transfection of siRNA to block PRL expression resulted in stimulation of retinal angiogenesis and vasodilation [13]. Finally, intravitreal administration of anti-PRL antibodies in neonatal rats caused a significant reduction in the number of vascular cells undergoing apoptosis in hyaloid vessels, a finding consistent with vasoinhibins promoting vascular regression during retinal development [21]. The potential importance of intraocular vasoinhibins is underscored by their presence in the eye of patients with advanced retinopathy of prematurity, where they are claimed to stimulate apoptosis-mediated vascular regression of newly formed intraocular blood vessels [21].

It is clear that vasoinhibins play a key role in preventing angiogenesis in the eye, and they may also maintain avascularity in cartilage. Chondrocytes express PRL mRNA and cleave the protein to vasoinhibins [8]. Also, chondrocytes are enriched with MMP that generate vasoinhibins [8]. Our current studies are addressing the role of vasoinhibins in the control of cartilage avascularity and their alteration in rheumatoid arthritis, an angiogenesis-dependent disease.

Vasoinhibins in the Pituitary Gland

In contrast to the above tissues, endocrine glands are highly vascular, and their blood supply is required to obtain the oxygen, nutrients and precursors necessary to synthesize large amounts of hormones. Furthermore, high permeability of local microvessels is essential for hormonal release into the bloodstream. Interest in the control of angiogenesis in the pituitary gland was raised by the observation that pituitary adenomas are less vascularized than normal tissue [43]. Because an adequate blood supply enables tumor cells to proliferate further or metastasize to another organ, the reduced vascularization of pituitary adenomas suggested that inhibition of angiogenesis in these tumors played a role in their characteristically slow growth and failure to reach the carcinoma state. Therefore, inhibitors of angiogenesis might influence the phenotype and behavior of these tumors.

Vasoinhibins have been detected in the normal [14] and tumorous [11] pituitary gland, but their role in pituitary angiogenesis remains to be elucidated. Of interest is the fact that hypoxia, a common feature of solid tumors and a main trigger of angiogenesis, decreases PRL synthesis and suppresses its cathepsin D-mediated conversion to vasoinhibins in the rat GH4C1 pituitary adenoma cell line [44]. Downregulation of vasoinhibins by hypoxia in pituitary tumor cells is consistent with the proposal that the angiogenic switch in highly vascular endocrine organs is achieved by reducing inhibitor production rather than by upregulating angiogenesis stimulators [43].

Conclusions and Perspectives

Vasoinhibins are natural inhibitors of blood vessel growth and dilation and promoters of vascular regression in the eye. The wide-spread expression of vasoinhibins and of the proteases (cathepsin-D and MMP) involved in their generation suggests that they are physiologically active at other sites as well. Because vasoinhibins need to be generated via proteolysis of PRL, regulation of the specific proteases is likely to play a key role in determining their actions under health and disease. The ability of vasoinhibins to influence both early and late events in the angiogenic cascade supports their being strong regulators of angiogenesis, and thus better targets to limit growth. However, further studies are needed to evaluate the underlying mechanisms of vasoinhibin actions, their receptors, and signaling pathways. Likewise, these peptides need to be fully characterized chemically in order to search for a structural homolog with perhaps more potent vasoinhibitory activity. This information is important for determining their role in normal and pathological processes and to determine if and how they could be applied therapeutically.

Acknowledgments

The authors thank F. López-Barrera and G. Nava for their expert technical assistance, and D.D. Pless for critically editing the manuscript. Grant support: Supported by National Autonomous University of Mexico (Grant IN224405) and the National Council of Science and Technology of Mexico (Grants SALUD-2004–C02–16 and 43401).

References

1 Cao Y, Cao R, Veitonmäki N: Kringle structures and antiangiogenesis. Curr Med Chem Anti-Canc Agents 2002;2:667–681.
2 Nyberg P, Xie L, Kalluri R: Endogenous inhibitors of angiogenesis. Cancer Res 2005;65: 3967–3979.
3 Ferrara N, Clapp C, Weiner R: The 16K fragment of prolactin specifically inhibits basal or fibroblast growth factor stimulated growth of capillary endothelial cells. Endocrinology 1991;129: 896–900.

4 Baldocchi RA, Tan L, King DS, Nicoll CS: Mass spectrometric analysis of the fragments produced by cleavage enzyme and reduction of rat prolactin: evidence that the cleaving enzyme is cathepsin D. Endocrinology 1993;133:935–938.
5 Clapp C, Martial J, Guzmán R, Rentier-Delrue F, Weiner R: The 16 kilodalton N-terminal fragment of human prolactin is a potent inhibitor of angiogenesis. Endocrinology 1993;133: 1292–1299.
6 Struman I, Bentzien F, Lee H, Mainfroid V, D'Angelo G, Goffin V, Weiner RI, Martial JA: Opposing actions of intact and N-terminal fragments of the human prolactin/growth hormone family members on angiogenesis: an efficient mechanism for the regulation of angiogenesis. Proc Natl Acad Sci 1999;96:1246–1251.
7 Tabruyn SP, Sorlet CM, Rentier-Delrue F, Bours V, Weiner RI, Martial JA, Struman I: The antiangiogenic factor 16K human prolactin induces caspase-dependent apoptosis by a mechanism that requires activation of nuclear factor-kappa B. Mol Endocrinol 2003;17:1815–1823.
8 Macotela Y, Aguilar MB, Guzmán-Morales J, Zermeño C, López-Barrera F, Nava G, Lavalle C, Martinez de la Escalera G, Clapp C: Matrix metalloproteases from chondrocytes generate antiangiogenic 16K-prolactin. J Cell Sci 2006;119:1790–1800.
9 Khurana S, Liby K, Bucley AR, Ben-Jonathan N: Proteolysis of human prolactin: resistance to cathepsin D and formation of a nonangiostatic, C-terminal 16K fragment by thrombin. Endocrinology 1999;140:4127–4132.
10 Keeler C, Dannies PS, Hodsdon ME: The tertiary structure and backbone dynamics of human prolactin. J Mol Biol 2003;328:1105–1121.
11 Piwnica D, Touraine P, Struman I, Tabruyn S, Bolbach G, Clapp C, Martial JA, Kelly PA, Goffin V: Cathepsin D processes human prolactin into multiple 16K-like N-terminal fragments: study of their antiangiogenic properties and physiological relevance. Mol Endocrinol 2004;18:2522–2542.
12 Corbacho AM, Martínez de la Escalera G, Clapp C: Roles of prolactin and related members of the prolactin/growth hormone/placental lactogen family in angiogenesis. J Endocrinol 2002;173: 219–238.
13 Aranda J, Rivera JC, Jeziorski MC, Riesgo-Escovar J, Nava G, López-Barrera F, Quiróz-Mercado H, Berger P, Martínez de la Escalera G, Clapp C: Prolactins are natural inhibitors of angiogenesis in the retina. Invest Ophthalmol Vis Sci 2005;46:2947–2953.
14 Torner L, Mejía S, López-Gómez FJ, Quintanar A, Martínez de la Escalera G, Clapp C: A 14-kilodalton prolactin-like fragment is secreted by the hypothalamo-neurohypophyseal system of the rat. Endocrinology 1995;136:5454–5460.
15 Clapp C, Torner L, Gutiérrez-Ospina G, Alcántara E, López-Gómez FJ, Nagano M, Kelly PA, Mejía S, Morales MA, Martínez de la Escalera G: The prolactin gene is expressed in the hypothalamic-neurohypophyseal system and the prolactin is processed into a 14-kDa fragment with activity like 16-kDa prolactin. Proc Natl Acad Sci 1994;91:10384–10388.
16 Torner L, Maloumby R, Nava G, Aranda J, Clapp C, Neumann ID: In vivo release and gene upregulation of brain prolactin in response to physiological stimuli. Eur J Neurosci 2004;19:1601–1608.
17 Corbacho AM, Macotela Y, Nava G, Torner L, Dueñas Z, Noris G, Morales MA, Martinez de la Escalera G, Clapp C: Human umbilical vein endothelial cells express multiple prolactin isoforms. J Endocrinol 2000a;166:53–62.
18 Corbacho AM, Nava G, Eiserich JP, Noris G, Macotela Y, Struman I, Martínez de la Escalera G, Clapp C: Proteolytic cleavage confers nitric oxide synthase inducing activity upon prolactin. J Biol Chem 2000b;275:13183–13186.
19 Warner MD, Sinha YN, Peabody CA: Growth hormone and prolactin variants in normal subjects: relative proportions in morning and afternoon samples. Horm Metab Res 1993;5:425–429.
20 Fukuoka H, Hamamoto R, Higurashi M: Heterogeneity of serum and amniotic fluid prolactin in humans. Horm Res 1991;35:58–63.
21 Dueñas Z, Rivera JC, Quiroz-Mercado H, Aranda J, Macotela Y, Montes de Oca P, López-Barrera F, Nava G, Guerrero JL, Suarez A, De Regil M, Martínez de la Escalera G, Clapp C: Prolactin in eyes of patients with retinopathy of prematurity: implications for vascular regression. Invest Ophthalmol Vis Sci 2004;45:2049–2055.
22 Saint-Geniez M, D'Amore PA: Development and pathology of the hyaloid choroidal and retinal vasculature. Int J Dev Biol 2004;48:1045–1058.

23 Fraser HM, Wulff C: Angiogenesis in the corpus luteum. Reprod Biol Endocrinol 2003;I:88.
24 Giatromanolaki A, Sivridis E, Koukourakis MI: Tumour angiogenesis: vascular growth and survival. APMIS 2004;112:431–440.
25 González C, Corbacho AM, Eiserich JP, García C, Lopez-Barrera F, Morales-Tlalpan V, Barajas-Espinosa A, Diaz-Muñoz M, Rubio R, Lin SH, Martinez de la Escalera G, Clapp C: 16K-prolactin inhibits activation of endothelial nitric oxide synthase, intracellular calcium mobilization, and endothelium-dependent vasorelaxation. Endocrinology 2004;145:5714–5722.
26 Tabruyn SP, Nguyen NQ, Cornet AM, Martial JA, Struman I: The antiangiogenic factor, 16-kDa human prolactin, induces endothelial cell cycle arrest by acting at both the G_0-G_1 and the G_2-M phases. Mol Endocrinol 2005;19:1932–1942.
27 Lee H, Struman I, Clapp C, Martial J, Weiner R: Inhibition of urokinase activity by the antiangiogenic factor 16K prolactin: activation of plasminogen activator inhibitor 1 expression. Endocrinology 1998;139:3696–3703.
28 Martini JF, Piot C, Humeau LM, Struman I, Martial J, Weiner R: The anti-angiogenic factor 16K PRL induces programmed cell death in endothelial cells by caspase activation. Mol Endocrinol 2000;14:1536–1549.
29 D'Angelo G, Struman I, Martial J, Weiner RI: Activation of mitogen-activated protein kinases by vascular endothleial growth factor and basic fibroblast growth factor in capillary endothelial cells is inhibited by antiangiogenic factor 16-kDa N-terminal fragment of prolactin. Proc Natl Acad Sci USA 1995;92:6374–6378.
30 Lee SH, Nishino M, Mazumdar T, Garcia GE, Galfione M, Lee FL, Lee CL, Liang A, Kim J, Feng L, Eissa NT, Lin SH, Yu-Lee LY: 16-kDa prolactin down-regulates inducible nitric oxide synthase expression through inhibition of the signal transducer and activator of transcription 1/IFN regulatory-pathway. Cancer Res 2005;65:7984–7992.
31 Macotela Y, Mendoza C, Corbacho AM, Cosio G, Eiserich JP, Zentella A, Martinez de la Escalera G, Clapp C: 16K prolactin induces NF-kappaB activation in pulmonary fibroblasts. J Endocrinol 2002;175:R13–R18.
32 Clapp C, Weiner RI: A specific, high affinity, saturable binding site for the 16-kilodalton fragment of prolactin on capillary endothelial cells. Endocrinology 1992;130:1380–1386.
33 Walford G, Loscalzo J: Nitric oxide in vascular biology. J Thromb Haemost 2003;1:2112–2118.
34 Montes de Oca P, Macotela Y, Nava G, Lopez-Barrera F, Martinez de la Escalera G, Clapp C: Prolactin stimulates integrin-mediated adhesion of circulating mononuclear cells to endothelial cells. Lab Invest 2005;85:633–642.
35 Horseman ND, Zhao W, Montecino-Rodriguez E, Tanaka M, Nakashima K, Engle SJ, Smith F, Markoff E, Dorshkind K: Defective mammopoiesis but normal hematopoyesis in mice with a targeted disruption of the prolactin gene. EMBO J 1997;16:6926–6935.
36 Bole-Feysot C, Goffin V, Edery M, Binart N, Kelly PA: Prolactin (PRL) and its receptor: actions, signal transduction pathways and phenotypes observed in PRL receptor knockout mice. Endocr Rev 1998;19:225–298.
37 Das A, McGuire P: Retinal and choroidal angiogenesis: pathophysiology and strategies for inhibition. Prog Retin Eye Res 2003;22:721–748.
38 Pleyer U, Gupta D, Weidle EG, Lisch W, Zierhut M, Thiel HJ: Elevated prolactin levels in human aqueous humor of patients with anterior uveitis. Graefes Arch Clin Exp Ophthalmol 1991;229: 447–451.
39 O'Steen WK, Sundberg DK: Patterns of radioactivity in the eyes of rats after injection of iodinated prolactin. Ophthalmic Res 1982;14:54–62.
40 Ochoa A, Montes de Oca P, Rivera JC, Dueñas Z, Nava G, de la Escalera GM, Clapp C: Expression of prolactin gene and secretion of prolactin by rat retinal capillary endothelial cells. Invest Ophthalmol Vis Sci 2001;42:1639–1645.
41 Pan H, Nguyen NQ, Yoshida H, Bentzien F, Shaw LC, Rentier-Delrue F, Martial JA, Weiner R, Struman I, Grant MB: Molecular targeting of antiangiogenic factor 16K hPRL inhibits oxygen-induced retinopathy in mice. Invest Ophthalmol Vis Sci 2004;45:2413–2419.
42 Dueñas Z, Torner L, Corbacho AM, Ochoa A, Gutierrez-Ospina G, Lopez-Barrera F, Barrios FA, Berger P, Martinez de la Escalera G, Clapp C: Inhibition of rat corneal angiogenesis by 16-kDa

prolactin and by endogenous prolactin like molecules. Invest Ophthalmol Vis Sci 1999;40: 2498–2505.
43 Turner HE, Harris AL, Melmed S, Wass JA: Angiogenesis in endocrine tumors. Endocr Rev 2003;24: 600–632.
44 Cosio G, Jeziorski MC, Lopez-Barrera F, De la Escalera GM, Clapp C: Hypoxia inhibits expresion of prolactin and secretion of cathepsin-D by the GH4C1 pituitary adenoma cell line. Lab Invest 2003;83:1627–1636.

Dr. Carmen Clapp
Instituto de Neurobiología, Universidad Nacional Autónoma de México
Campus UNAM-Juriquilla, Querétaro, Qro. 76230 (México)
Tel. +52 4422381028, Fax +52 4422340344, E-Mail clapp@servidor.unam.mx

Arzt E, Bronstein M, Guitelman M (eds): Pituitary Today: Molecular, Physiological and Clinical Aspects. Front Horm Res. Basel, Karger, 2006, vol 35, pp 74–87

Molecular Mechanisms of Pituitary Differentiation and Regulation: Implications for Hormone Deficiencies and Hormone Resistance Syndromes

Jacques Drouin

Laboratoire de Génétique Moléculaire, Institut de Recherches Cliniques de Montréal (IRCM), Montréal, Canada

Abstract

During the last century, the questions posed by scientists and clinicians on pituitary function have led to new concepts about the mechanisms of hormone action and cell differentiation. In particular, the advent of molecular genetics and the cloning of pituitary hormone coding genes followed by discovery of their regulators during the last 20 years has provided tremendous insight into the pathophysiological bases of hormone deficit and excess, as well as offering novel therapeutic opportunities. Most insight was gained through the identification of transcription factors that control the program of pituitary organogenesis and cell differentiation; it is indeed the normal developmental program controlled by these transcriptional regulators that is perturbed in inherited forms of hormone deficiency. This review will summarize our current understanding of these processes and their implications for hormone deficiency and hormone resistance syndromes from a developmental perspective.

Formation of Rathke's Pouch and Pituitary

Rathke's pouch, the pituitary anlage, is formed as the developing neuroepithelium is separated from the surface ectoderm by invasion of prechordal mesoderm and neural crest derivatives. During craniofacial development, the separation of these two epithelia (neuroepithelium that will become the brain and surface ectoderm that will become the oral epithelium or stomodeum) occurs everywhere except in the midline region that will become the back of the oral cavity. Maintenance of a tight contact between neuroepithelium and surface ectoderm thus forms Rathke's pouch: this invagination is not the consequence of an active process but rather formed passively through separation of two epithelial layers. Rathke's pouch will then separate for the oral epithelium and form the original glandular structure that is constituted of an epithelium surrounding a lumen. Cells then leave this epithelium on its ventral side and proliferate to form the nascent anterior lobe where cell differentiation is initiated.

Although the developmental origins of the pituitary have been most clearly defined in the chick/quail system [1, 2], molecular genetic determinants of this process have been identified and studied in the mouse. The earliest factors that impact on pituitary development and mark its developmental origins are the Pitx1 (Ptx1) and Pitx2 transcription factors. Both factors are expressed in the stomodeum and their expression is maintained in the developing pituitary and in the adult gland [3–6]. The founding member of this family, Pitx1, was originally cloned because of its implication in cell-specific transcription of the POMC gene [7], hence the name of this subfamily of homeodomain transcription factors that otherwise play significant roles in early development, Pitx1 being the major determinant for hindlimb identity [8] and Pitx2 being an effector of left-right asymmetry during development of internal organs [9–12]. In agreement with the idea that *Pitx1* and *Pitx2* are molecular signatures of the embryological origin of the pituitary (fig. 1), their inactivation causes craniofacial malformations and leads to interruption of pituitary development. While the knockout of *Pitx1* has the least severe phenotype with less gonadotropic and thyrotropic cells [8, 13] in agreement with the higher levels of *Pitx1* protein in these cells [4], the knockout of *Pitx2* leads to arrested pituitary development at the late Rathke's pouch stage [10, 11, 14]. However, it is the double mutant for *Pitx1* and *Pitx2* that clearly reveals the importance of Pitx factors in early pituitary development as the double knockout leads to arrest at the early Rathke's pouch stage; this also prevents expression of the downstream *Lhx3* gene [15]. These gene mutations have clearly indicated that the *Pitx* genes are required for pituitary organogenesis, in particular for the transition from Rathke's pouch to glandular structure, for the initiation of cell proliferation and for the onset of the cell differentiation program (fig. 1).

The related *Lhx3* and *Lhx4* genes are also expressed at the midline in the oral ectoderm but their onset of expression is later than the Pitx factors. As indicated above, the *Pitx1*−/−; *Pitx2*−/− embryos fail to switch-on *Lhx3* expression, clearly suggesting that the later is downstream of the Pitx factors in the regulatory cascade controlling early pituitary organogenesis [15, 16]. In fact, this interpretation is consistent with the phenotype of knockout mice as *Lhx3*−/− embryos have arrested pituitary development at the late Rathke's pouch stage, a phenotype that is quite similar to that of *Pitx2*−/− mice [11, 17]. Whereas the *Lhx4*−/− mice have a hypoplasic anterior pituitary [18], mice that

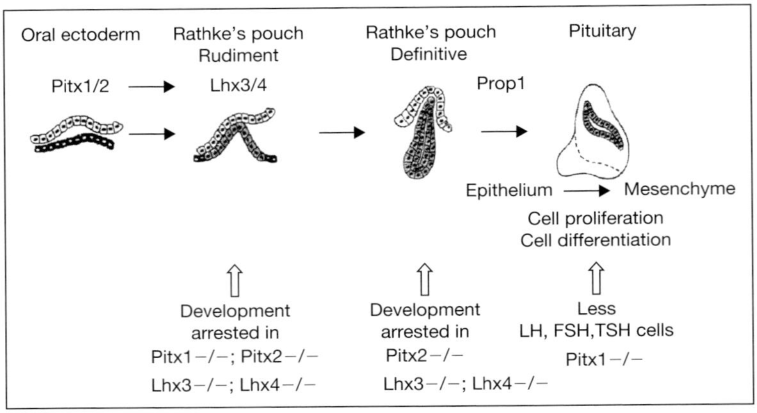

Fig. 1. Formation of Rathke's pouch and early pituitary development.

are deficient in both *Lhx3* and *Lhx4* have arrested development at the early pouch stage and in that sense, they appear to be a phenocopy of the *Pitx1*−/−; *Pitx2*−/− phenotype [15, 18].

Epithelium-Mesenchyme Transition

When Rathke's pouch separates from the oral ectoderm, it initially forms a closed epithelial structure with a lumen. Soon after, cells come out of the ventral side of this epithelium and take a mesenchymal appearance as they proliferate: this is closely associated with the initiation of the cell differentiation program (fig. 1). The homeodomain transcription factor Prop1 (for Prophet-of-Pit1) [19] appears to be important for the transition between epithelium to mesenchyme as *Prop1*−/− mice exhibit an expanded epithelium that first appears to proliferate and later regresses through apoptosis to yield a hypoplasic pituitary [20]. This leads to a complex phenotype in terms of hormone secreting cells with the most notable being the deficiency in GH, PRL and TSH that results from absence of Pit1 as indicated in the name Prop1. It was later found that Prop1 deficiency may also lead to complete or partial deficiencies of LH/FSH in agreement with inappropriate SF1 expression; this is a likely cause of delayed puberty and hypogonadism associated with this mutation. In addition, *Prop1*-deficient mice have altered POMC expression [21]. The action of Prop1 may be balanced by another homeodomain transcription factor, Hesx1 (Rpx) that is expressed in Rathke's pouch and in the adjoining ventral diencephalon [22].

Table 1. Combined pituitary hormone deficiency

Deficiency	Affected gene	Phenotype
GH or many	HESX1 (RPX)	septo-optic dysplasia (SOD)
GH, PRL, TSH, LH/FSH	LHX3	hypopituitarism
GH, TSH, ACTH	LHX4	hypopituitarism
GH, TSH, LH/FSH, ACTH	PROP-1	hypopituitarism
GH, PRL, TSH	PIT-1	hypopituitarism

Hesx1 (Rpx) was shown to be a repressor of Prop1 activity [19] and Prop1 contributes to extinction of *Hesx* expression [23, 24].

Collectively, these genes play critical roles in the early steps of pituitary organogenesis and it is thus not surprising that some of them are involved in multiple hormone deficiencies.

Combined Pituitary Hormone Deficiencies

Given their role in pituitary organogenesis as revealed in mouse studies, it is not surprising that human mutations in genes *LHX3*, *LHX4*, *HESX1* and *PROP1* have been associated with combined pituitary hormone deficiencies (CPHD; table 1). Actually, *HESX1* mutations have been associated with rare cases of septo-optic-dysplasia (SOD) that include cranio-malformations in addition to pituitary hypoplasia and disruption of the pituitary stalk [25, 26]. Mutations of either *Lhx3* or *Lhx4* have also been associated with CPHD but this is quite rare as only a few families have been identified [27, 28]. While mutations of *PITX1* have not yet been identified in humans, haploinsufficiency for *PITX2* causes Rieger syndrome and this is occasionally associated with GH deficiency [5].

The most frequent cause of CPHD is, however, due to PROP1 mutations. Most of these mutations affect the homeodomain coding region of *PROP1* [29]. *PROP1*-dependent CPHD is recessive and it usually involves complete deficiency of GH, TSH and LH/FSH with variable progression of the disease and about half the patients eventually developing ACTH deficiency [30, 31]. Interestingly, *PROP1*-deficient CPHD patients have been variably found to have at MRI either hyper- or hypoplasia of the pituitary in agreement with the sequential expansion of the pituitary epithelium and subsequent tissue loss observed in *Prop1*-deficient mice [20].

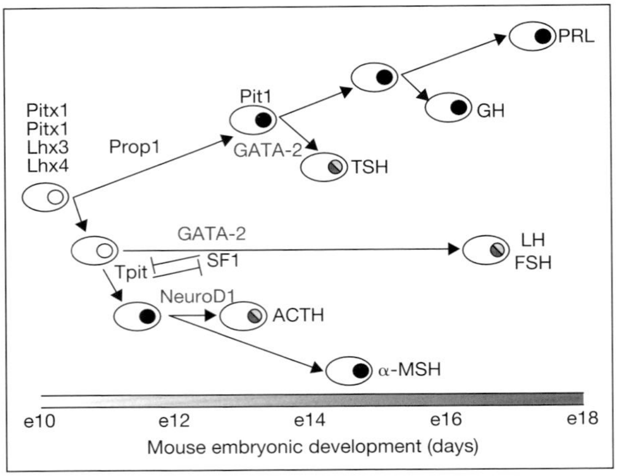

Fig. 2. Pituitary cell differentiation: a binary model.

It thus appears that human CPHD is quite similar to the mouse models of this condition and that further investigation of the mouse models may enlighten us on disease progression and identification of other causative genes.

Sequential Pituitary Cell Differentiation: A Binary Model

All cells of the anterior and intermediate pituitary lobes derive from Rathke's pouch and although it is quite possible that commitment occurs early in various regions of the pouch, the first visible markers of cell differentiation appear at e12-e12.5 in the mouse at the same time as the onset of transition from epithelium to mesenchyme. The first terminally differentiated cells to be detected are corticotropes marked by Tpit expression and these first cells appear in the epithelial part of the early gland [32]. This is followed closely by the first thyrotropes and by another Pit1-dependent lineage, the somatotropes. At around e14.5, the first evidence of terminal differentiation of melanotropes is detected in the intermediate lobe and the gonadotropes are the last to reach terminal differentiation around e16.5 [33].

The relationships between various lineages were defined by the investigation of mutations in the genes encoding cell-restricted transcription factors that control not only transcription of hormone-coding genes but also differentiation of the corresponding lineage (fig. 2). Thus, analysis of Pit1 mutant mice defined the relationships between somatotropic, lactotropic and thyrotropic cells [34] whereas it is essentially the analysis of *Tpit−/−* mice that reveal the close relationship between corticotropes/melanotropes and gonadotropes [35]. This work indicated that Tpit is a positive regulator for corticotropes/melanotropes and that it is a negative regulator of the gonadotropic cell fate, thus clearly suggesting that these lineages have a common precursor. This precursor and the lineages that derive from it are clearly separate from the Pit1-dependent lineages that constitute the other half of a differentiation sequence that can be accounted for by a series of a binary cell fate choices starting with a putative pituitary precursor [35, 36]. Although Tpit and Pit1 constitute critical regulators of cell fate, it appears that lineage commitment is likely to occur upstream of these factors by yet unknown mechanisms and factors.

Pit1-Dependent Lineages

Pit1 (GHF1) is a POU homeodomain transcription factor that was originally cloned for its role in GH transcription [37, 38]. It is expressed from about e13.5 in GH, PRL and TSH cells, and when it is mutated, it causes deficiencies in the corresponding hormones and pituitary hypoplasia [34]. The Snell and Jackson dwarf mice harbor mutations in this gene. Pit1 expression is dependent on Prop1 that was itself found to be the causative gene for dwarfism in Ames mice [19, 20].

It is noteworthy that Pitx1 is critical for expression of the three Pit-dependent hormone genes and that this occurs through interactions with Pit1 [39, 40]. The dependence on Pit1 is taken to indicate that the three Pit-dependent lineages have a common precursor. However, the mechanisms/regulators that define each of those lineages remain unknown: GATA-2 may play a role in the differentiation of thyrotropes [41] and the estrogen receptor is likely to play an important role in postnatal PRL expression [42].

Gonadotropes

Gonadotropins are heterodimers between a common α-glycoprotein subunit (α-GSU) and a hormone-specific β subunit. As for all pituitary hormone-coding genes, Pitx1 plays a role in transcription of the *αGSU*, *βLH* and *βFSH* genes [16, 43–47]. This action is exerted in cooperation with the orphan nuclear SF1 that is expressed in steroidogenic tissues, adrenals and gonads, in hypothalamic GnRH neurons and in pituitary gonadotropes. *SF1*−/− mice do not express gonadotropins and have adrenal and gonadal agenesis together with deficient development of hypothalamic GnRH neurons [48]. Treatment of *SF1*−/− mice with GnRH leads to activation of gonadotropin gene expression suggesting that SF1 is a relatively late regulator of gonadotrope differentiation [49]. GnRH signaling in gonadotropes also requires Egr1 that acts synergistically with Pitx1 for transcription of the βLH gene [44].

Finally, GATA-2 appears to be important as a positive regulator of gonadotrope differentiation as suggested in gain-of-function experiments [41]. The exact role of GATA-2 is puzzling as this factor is expressed much earlier than differentiation of gonadotropes and thus its effect on gonadotrope differentiation may not be direct.

Corticotropes and Melanotropes: Two POMC-Expressing Lineages

One transcription factor is critical for differentiation of the two POMC lineages, the Tbox factor Tpit. Tpit is highly restricted in its expression to those two lineages where it is detected about 12 h before POMC [32]. Knockout of the *Tpit* gene causes almost complete depletion of POMC-expressing corticotropes and melanotropes although a few positive cells remain in both anterior and intermediate lobes [34]. The *Tpit* loss-of-function mice clearly supported the role of Tpit in transcription of the POMC gene and showed that it is essential for either expansion or maintenance of corticotropes and melanotropes; however, it also suggested that the initial commitment to these lineages is independent of Tpit and presumably triggered by other upstream mechanisms and factors [34].

On the POMC promoter, Tpit exerts it action by close cooperative physical interactions with Pitx1 [32] and the Tpit/Pitx factors exhibit physical and transcriptional synergism with bHLH heterodimers containing the neurogenic bHLH NeuroD1 [50, 51]. NeuroD1 is expressed in corticotropes and clearly excluded from intermediate lobe melanotropes: it is thus one candidate factor to differentiate the two POMC-expressing lineages. The *NeuroD1* knockout only has a minimal effect on differentiation of corticotropes as it appears to delay the appearance of POMC in corticotropes by about 2–3 days [51].

Expression of POMC is stimulated by the hypothalamic hormone CRH and this action appears to be mediated through the PKA and MAP kinase pathways [52, 53] and to ultimately stimulate the activity of NGFI-B and the related orphan nuclear receptors [54, 55] as well as through Tpit and the recruitment of the SRC2 coactivator [56]. LIF also stimulates POMC transcription in synergism with CRH and this action is mediated by STAT3 [57, 58].

An important aspect of the hypothalamo-pituitary-adrenal (HPA) axis homeostasis is the negative feedback regulation exerted by glucocorticoids on corticotrope POMC expression. This negative feedback is mediated by the glucocorticoid receptor on both secretion of pre-stored ACTH and on transcription of the POMC gene [59]. Repression of POMC transcription by the glucocorticoid receptor may be exerted in part through binding of the negative glucocorticoid response element (nGRE) present at −63 bp of the mouse POMC promoter [60] but it also involves trans-repression of NGFI-B-dependent transcription [61, 62]. Trans-repression is the mutual antagonism exerted between two transcription factors such as GR and NGFI-B, and it results from direct protein:protein interactions between these transcriptional activators; it is a reciprocal mechanism such that the DNA-bound factor behaves as transcriptional activator whereas the factor recruited through protein:protein interaction acts as repressor. In the case of POMC, NGFI-B is bound to the POMC promoter NurRE and this recruitment is enhanced by CRH signaling [61] and GR recruited to promoter-bound NGFI-B antagonizes its transcriptional activity [62]. Recent work has identified proteins that stabilize these interactions in corticotropes and that may be so important as to prevent trans-repression of NGFI-B activity by GR in their absence. Loss of expression of this cofactor causes glucocorticoid resistance and may account for this resistance in corticotropic adenomas found in patients with Cushing disease.

Isolated Pituitary Hormone Deficiencies

Most cases of isolated pituitary hormone deficiencies have been associated with mutations in the hormone-coding genes themselves (e.g. in GR, TSH-β, FSH-β, LH-β and POMC) or in the genes encoding receptors for hypothalamic hormones (such as GHRH-R, TRH-R and GnRH-R). A few cases have been associated with mutations in transcription factors, such as SF1 and DAX-1 that are associated with hypogonadotropic hypogonadism as well as with adrenal malformations (table 2).

Isolated ACTH Deficiency

Isolated ACTH deficiency (IAD) was thought like other isolated hormone deficiencies to be very rare. Indeed, only 4 case reports suggested that this condition may exist [63–66]. Human POMC gene mutations cause pituitary ACTH deficiency but also obesity and hypopigmentation (red hair) in agreement with expression of the POMC in the pituitary, hypothalamic neurons and skin [67]. Since Tpit expression is restricted to pituitary corticotropes, we postulated that mutations in this gene could cause IAD and this hypothesis was confirmed in IAD patients [32, 68, 69]. After we initially found a Tpit mutation in one of the early case reports of IAD [32, 63], we investigated and found other Tpit mutations in many similar cases of this condition that now appears to be clinically very homogeneous [69]. Tpit-dependent IAD typically has an early neonatal onset and can be fatal if the severe hypoglycemic episodes that always lead to

Table 2. Isolated pituitary hormone deficiencies

Deficiency	Affected gene	Phenotype
GH	GH GHRH-R	dwarfism
TSH	TSH-β TRH-R	central hypothyroidism
LH/FSH	FSH-β, LH-β GnRH-R KAL-1 SF1 DAX-1 Leptin, leptin-R	hypogonadotropic hypogonadism hypogonadotropic hypogonadism + anosmia (Kallmann syndrome) + sex reversal and adrenal failure + congenital adrenal hypoplasia + obesity
ACTH	POMC TPIT	adrenal insufficiency, obesity, red hair adrenal insufficiency

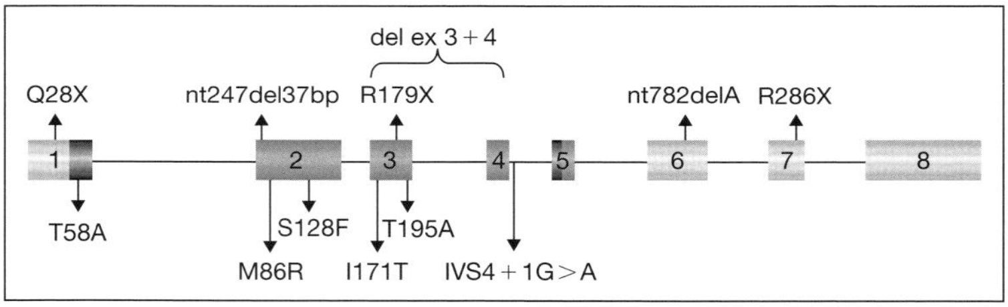

Fig. 3. *TPIT* mutations identified in neonatal onset isolated ACTH deficiency (IAD).

diagnosis are not treated rapidly. We have now identified a variety of loss-of-function mutations in the Tpit gene that cause IAD (fig. 3): these include insertion of premature stop codons that lead to mRNA degradation by non-sense-mediated mRNA decay, amino acid replacements that impair the DNA binding ability of Tpit and splicing mutations.

Perspectives

The genetic analysis of pituitary development has increased dramatically our understanding of the complex mechanisms controlling organogenesis, cell differentiation and hormone action. However, this sense of accomplishment must be balanced by the realization that we are still ignorant of many regulators for early lineage commitment, terminal differentiation of several lineages and for integration of the multiple hormonal signaling pathways that fine tune the production of each pituitary hormone. The successes of the last decade indicate that the discovery of those unknown regulators and their mechanisms of action will have a direct impact not only for understanding, but also for treatment of pituitary disorders: it is therefore our challenge to uncover those regulatory mechanisms/factors and their implications in the pathophysiology of pituitary diseases.

Acknowledgments

We thank the numerous colleagues and members of the laboratory who have made our contribution to the field possible. Work in the author's laboratory has been funded by the National Cancer Institute of Canada and the Canadian Institutes of Health Research (CIHR).

References

1 Couly GF, Le Douarin NM: Mapping of the early neural primordium in quail-chick chimeras. I. Developmental relationships between placodes, facial ectoderm, and prosencephalon. Dev Biol 1985;110:422–439.
2 Couly GF, Le Douarin NM: Mapping of the early neural primordium in quail-chick chimeras. II. The prosencephalic neural plate and neural folds: implications for the genesis of cephalic human congenital abnormalities. Dev Biol 1987;120:198–214.
3 Lanctôt C, Lamolet B, Drouin J: The *bicoid*-related homeoprotein Ptx1 defines the most anterior domain of the embryo and differentiates posterior from anterior lateral mesoderm. Development 1997;124:2807–2817.
4 Lanctôt C, Gauthier Y, Drouin J: Pituitary homeobox 1 (Ptx1) is differentially expressed during pituitary development. Endocrinology 1999;140:1416–1422.
5 Semina EV, Reiter R, Leysens NJ, Alward WL, Small KW, Datson NA, Siegel-Bartelt J, Bierke-Nelson D, Bitoun P, Zabel BU, Carey JC, Murray JC: Cloning and characterization of a novel bicoid-related homeobox transcription factor gene, RIEG, involved in Rieger syndrome. Nat Genet 1996;14:392–399.
6 Mucchielli ML, Martinez S, Pattyn A, Goridis C, Brunet JF: *Otlx2*, an *Otx*-related homeobox gene expressed in the pituitary gland and in a restricted pattern in the forebrain. Mol Cell Neurosci 1996;8:258–271.
7 Lamonerie T, Tremblay JJ, Lanctôt C, Therrien M, Gauthier Y, Drouin J: PTX1, a *bicoid*-related homeobox transcription factor involved in transcription of pro-opiomelanocortin (POMC) gene. Genes Dev 1996;10:1284–1295.
8 Lanctôt C, Moreau A, Chamberland M, Tremblay ML, Drouin J: Hindlimb patterning and mandible development require the *Ptx1* gene. Development 1999;126:1805–1810.
9 Lu MF, Pressman C, Dyer R, Johnson RL, Martin JF: Function of Rieger syndrome gene in left-right asymmetry and craniofacial development. Nature 1999;401:276–278.
10 Lin CR, Kioussi C, O'Connell S, Briata P, Szeto D, Liu R, Izpisua Belmonte JC, Rosenfeld MG: Pitx2 regulates lung asymmetry, cardiac positioning and pituitary and tooth morphogenesis. Nature 1999;401:279–282.

11 Gage PJ, Suh HY, Camper SA: Dosage requirement of Pitx2 for development of multiple organs. Development 1999;126:4643–4651.

12 Kitamura K, Miura H, Miyagawa-Tomita S, Yanazawa M, Katoh-Fukui Y, Suzuki R, Ohuchi H, Suehiro A, Motegi Y, Nakahara Y, Kondo S, Yokoyama M: Mouse Pitx2 deficiency leads to anomalies of the ventral body wall, heart, extra- and periocular mesoderm and right pulmonary isomerism. Development 1999;126:5749–5758.

13 Szeto DP, Rodriguez-Esteban C, Ryan AK, O'Connell SM, Liu F, Kioussi C, Gleiberman AS, Izpisua-Belmonte JC, Rosenfeld MG: Role of the bicoid-related homeodomain factor Pitx1 in specifying hindlimb morphogenesis and pituitary development. Genes Dev 1999;13:484–494.

14 Suh H, Gage PJ, Drouin J, Camper SA: *Pitx2* is required at multiple stages of pituitary organogenesis: pituitary primordium formation and cell specification. Development 2002;129:329–337.

15 Charles MA, Suh H, Drouin J, Camper SA, Gage PJ: PITX genes are required for cell survival and *Lhx3* activation. Mol Endocrinol 2005;19:1893–1903.

16 Tremblay JJ, Lanctôt C, Drouin J: The pan-pituitary activator of transcription, Ptx-1 (pituitary homeobox1), acts in synergy with SF-1 and Pit1 and is an upstream regulator of the Lim-homeodomain gene Lim3/Lhx3. Mol Endocrinol 1998;12:428–441.

17 Sheng HZ, Zhadanov AB, Mosinger B, Fujii T, Bertuzzi S, Grinberg A, Lee EJ, Huang SP, Mahon KA, Westphal H: Specification of pituitary cell lineages by the LIM homeobox gene Lhx3. Science 1996;272:1004–1007.

18 Sheng HZ, Moriyama K, Yamashita T, Li H, Potter SS, Mahon KA, Westphal H: Multistep control of pituitary organogenesis. Science 1997;278:1809–1812.

19 Sornson MW, Wu W, Dasen JS, Flynn SE, Norman DJ, O'Connell SM, Gukovsky I, Carriere C, Ryan AK, Miller AP, Zuo L, Gleiberman AS, Andersen B, Beamer WG, Rosenfeld MG: Pituitary lineage determination by the Prophet of Pit-1 homeodomain factor defective in Ames dwarfism. Nature 1996;384:327–333.

20 Ward RD, Raetzman LT, Suh H, Stone BM, Nasonkin IO, Camper SA: Role of PROP1 in pituitary gland growth. Mol Endocrinol 2005;19:698–710.

21 Raetzman LT, Ward R, Camper SA: Lhx4 and Prop1 are required for cell survival and expansion of the pituitary primordia. Development 2002;129:4229–4239.

22 Hermesz E, Williams-Simons L, Mahon KA: A novel inducible element, activated by contact with Rathke's pouch, is present in the regulatory region of the Rpx/Hesx1 homeobox gene. Dev Biol 2003;260:68–78.

23 Gage PJ, Roller ML, Saunders TL, Scarlett LM, Camper SA: Anterior pituitary cells defective in the cell-autonomous factor, d.f., undergo cell lineage specification but not expansion. Development 1996;122:151–160.

24 Dasen JS, Barbera JP, Herman TS, Connell SO, lson LO, Ju B, Tollkuhn J, Baek SH, Rose DW, Rosenfeld MG: Temporal regulation of a paired-like homeodomain repressor/TLE corepressor complex and a related activator is required for pituitary organogenesis. Genes Dev 2001;15: 3193–3207.

25 Dattani MT, Martinez-Barbera JP, Thomas PQ, Brickman JM, Gupta R, Martensson IL, Toresson H, Fox M, Wales JK, Hindmarsh PC, Krauss S, Beddington RS, Robinson IC: Mutations in the homeobox gene HESX1/Hesx1 associated with septo-optic dysplasia in human and mouse. Nat Genet 1998;19:125–133.

26 Dattani MT: Growth hormone deficiency and combined pituitary hormone deficiency: does the genotype matter? Clin Endocrinol (Oxf) 2005;63:121–130.

27 Netchine I, Sobrier ML, Krude H, Schnabel D, Maghnie M, Marcos E, Duriez B, Cacheux V, Moers A, Goossens M, Gruters A, Amselem S: Mutations in LHX3 result in a new syndrome revealed by combined pituitary hormone deficiency. Nat Genet 2000;25:182–186.

28 Machinis K, Pantel J, Netchine I, Leger J, Camand OJ, Sobrier ML, Dastot-Le Moal F, Duquesnoy P, Abitbol M, Czernichow P, Amselem S: Syndromic short stature in patients with a germline mutation in the LIM homeobox LHX4. Am J Hum Genet 2001;69:961–968.

29 Cohen LE, Radovick S: Molecular basis of combined pituitary hormone deficiencies. Endocr Rev 2002;23:431–442.

30 Vallette-Kasic S, Barlier A, Teinturier C, Diaz A, Manavela M, Berthezene F, Bouchard P, Chaussain JL, Brauner R, Pellegrini-Bouiller I, Jaquet P, Enjalbert A, Brue T: PROP1 gene

screening in patients with multiple pituitary hormone deficiency reveals two sites of hypermutability and a high incidence of corticotroph deficiency. J Clin Endocrinol Metab 2001;86: 4529–4535.
31 Bottner A, Keller E, Kratzsch J, Stobbe H, Weigel JF, Keller A, Hirsch W, Kiess W, Blum WF, Pfaffle RW: PROP1 mutations cause progressive deterioration of anterior pituitary function including adrenal insufficiency: a longitudinal analysis. J Clin Endocrinol Metab 2004;89: 5256–5265.
32 Lamolet B, Pulichino AM, Lamonerie T, Gauthier Y, Brue T, Enjalbert A, Drouin J: A pituitary cell-restricted T-box factor, Tpit, activates POMC transcription in cooperation with Pitx homeoproteins. Cell 2001;104:849–859.
33 Japon MA, Rubinstein M, Low MJ: In situ hybridization analysis of anterior pituitary hormone gene expression during fetal mouse development. J Histochem Cytochem 1994;42: 1117–1125.
34 Li S, Crenshaw EBI, Rawson EJ, Simmons DM, Swanson LW, Rosenfeld MG: Dwarf locus mutants lacking three pituitary cell types result from mutations in the POU-domain gene pit-1. Nature 1990;347:528–533.
35 Pulichino AM, Vallette-Kasic S, Tsai JPY, Couture C, Gauthier Y, Drouin J: Tpit determines alternate fates during pituitary cell differentiation. Genes Dev 2003b;17:738–747.
36 Pulichino AM, Vallette-Kasic S, Drouin J: Transcriptional regulation of pituitary gland development: binary choices for cell differentiation. Curr Opin Endocrinol Diabetes 2004;11: 13–17.
37 Ingraham HA, Chen R, Mangalam HJ, Elsholtz HP, Flynn SE, Lin CR, Simmons DM, Swanson L, Rosenfeld MG: A tissue-specific transcription factor containing a homeodomain specifies a pituitary phenotype. Cell 1988;55:519–529.
38 Bodner M, Castrillo JL, Theill LE, Deerinck T, Ellisman M, Karin M: The pituitary-specific transcription factor GHF-1 is a homeobox-containing protein. Cell 1988;55:505–518.
39 Szeto DP, Ryan AK, O'Connell SM, Rosenfeld MG: P-OTX: a PIT-1-interacting homeodomain factor expressed during anterior pituitary gland development. Proc Natl Acad Sci USA 1996;93: 7706–7710.
40 Quentien MH, Manfroid I, Moncet D, Gunz G, Muller M, Grino M, Enjalbert A, Pellegrini I: Pitx factors are involved in basal and hormone-regulated activity of the human prolactin promoter. J Biol Chem 2002;277:44408–44416.
41 Dasen JS, O'Connell SM, Flynn SE, Treier M, Gleiberman AS, Szeto DP, Hooshmand F, Aggarwal AK, Rosenfeld MG: Reciprocal interactions of Pit1 and GATA2 mediate signaling gradient-induced determination of pituitary cell types. Cell 1999;97:587–598.
42 Simmons DM, Voss JW, Ingraham HA, Holloway JM, Broide RS, Rosenfeld MG, Swanson LW: Pituitary cell phenotypes involve cell-specific Pit-1 mRNA translation and synergistic interactions with other classes of transcription factors. Genes Dev 1990;4:695–711.
43 Tremblay JJ, Drouin J: Egr-1 is a downstream effector of GnRH and synergizes by direct interaction with Ptx1 and SF-1 to enhance luteinizing hormone β gene transcription. Mol Cell Biol 1999;19:2567–2576.
44 Tremblay JJ, Marcil A, Gauthier Y, Drouin J: Ptx1 regulates SF-1 activity by an interaction that mimics the role of the ligand-binding domain. EMBO J 1999;18:3431–3441.
45 Quirk CC, Lozada KL, Keri RA, Nilson JH: A single Pitx1 binding site is essential for activity of the LHbeta promoter in transgenic mice. Mol Endocrinol 2001;15:734–746.
46 Suszko MI, Lo DJ, Suh H, Camper SA, Woodruff TK: Regulation of the rat follicle-stimulating hormone beta-subunit promoter by activin. Mol Endocrinol 2003;17:318–332.
47 Melamed P, Koh M, Preklathan P, Bei L, Hew C: Multiple mechanisms for Pitx-1 transactivation of a luteinizing hormone beta subunit gene. J Biol Chem 2002;277:26200–26207.
48 Luo X, Ikeda Y, Parker KL: A cell-specific nuclear receptor is essential for adrenal and gonadal development and sexual differentiation. Cell 1994;77:481–490.
49 Ikeda Y, Luo X, Abbud R, Nilson JH, Parker KL: The nuclear receptor steroidogenic factor 1 is essential for the formation of the ventromedial hypothalamic nucleus. Mol Endocrinol 1995;9: 478–486.
50 Poulin G, Lebel M, Chamberland M, Paradis FW, Drouin J: Specific protein:protein interaction between basic Helix-Loop-Helix transcription factors and homeoproteins of the Pitx family. Mol Cell Biol 2000;20:4826–4837.

51 Lamolet B, Poulin G, Chu K, Guillemot F, Tsai MJ, Drouin J: Tpit-independent function of NeuroD1(BETA2) in pituitary corticotroph differentiation. Mol Endocrinol 2004;18:995–1003.
52 Reisine T, Rougon G, Barbet J, Affoter HU: Corticotropin-releasing factor-induced adrenocorticotropin-hormone release and synthesis is blocked by incorporation of the inhibitor of cyclic AMP-dependent protein kinase into anterior pituitary tumor cells by liposomes. Proc Natl Acad Sci USA 1985;82:8261–8265.
53 Kovalovsky D, Refojo D, Liberman AC, Hochbaum D, Pereda MP, Coso OA, Stalla GK, Holsboer F, Arzt E: Activation and induction of NUR77/NURR1 in corticotrophs by CRH/cAMP: involvement of calcium, protein kinase A, and MAPK pathways. Mol Endocrinol 2002;16: 1638–1651.
54 Philips A, Lesage S, Gingras R, Maira MH, Gauthier Y, Hugo P, Drouin J: Novel dimeric Nur77 signaling mechanisms in endocrine and lymphoid cells. Mol Cell Biol 1997a;17:5946–5951.
55 Maira MH, Martens C, Philips A, Drouin J: Heterodimerization between members of the Nur subfamily of orphan nuclear receptors as a novel mechanism for gene activation. Mol Cell Biol 1999;19:7549–7557.
56 Maira MH, Couture C, Le Martelot G, Pulichino AM, Bilodeau S, Drouin J: The T-box factor Tpit recruits SRC/p160 coactivators and mediates hormone action. J Biol Chem 2003;278: 46523–46532.
57 Ray DW, Ren SG, Melmed S: Leukemia inhibitory factor (LIF) stimulates proopiomelanocortin (POMC) expression in a corticotroph cell line. Role of STAT pathway. J Clin Invest 1996;97: 1852–1859.
58 Mynard V, Latchoumanin O, Guignat L, Devin-Leclerc J, Bertagna X, Barre B, Fagart J, Coqueret O, Catelli MG: Synergistic signaling by corticotropin-releasing hormone and leukemia inhibitory factor bridged by phosphorylated 3′,5′-cyclic adenosine monophosphate response element binding protein at the Nur response element (NurRE)-signal transducers and activators of transcription (STAT) element of the proopiomelanocortin promoter. Mol Endocrinol 2004;18: 2997–3010.
59 Drouin J, Maira MH, Philips A: Novel mechanism of action for Nur77 and antagonism by glucocorticoids: a convergent mechanism for CRH activation and glucocorticoid repression of POMC gene transcription. J Steroid Biochem Mol Biol 1998;65:59–63.
60 Drouin J, Sun YL, Chamberland M, Gauthier Y, De Léan A, Nemer M, Schmidt TJ: Novel glucocorticoid receptor complex with DNA element of the hormone-repressed POMC gene. EMBO J 1993;12:145–156.
61 Philips A, Maira MH, Mullick A, Chamberland M, Lesage S, Hugo P, Drouin J: Antagonism between Nur77 and glucocorticoid receptor for control of transcription. Mol Cell Biol 1997b;17: 5952–5959.
62 Martens C, Bilodeau S, Maira M, Gauthier Y, Drouin J: Protein: protein interactions and transcriptional antagonism between the subfamily of NGFI-B/Nur77 orphan nuclear receptors and glucocorticoid receptor. Mol Endocrinol 2005;19:885–897.
63 Malpuech G, Vanlieferinghen P, Dechelotte P, Gaulme J, Labbe A, Guiot F: Isolated familial adrenocorticotropin deficiency: prenatal diagnosis by maternal plasma estriol assay. Am J Med Genet 1988;29:125–130.
64 Kyllo JH, Collins MM, Vetter KL, Cuttler L, Rosenfield RL, Donohoue PA: Linkage of congenital isolated adrenocorticotropic hormone deficiency to the corticotropin releasing hormone locus using simple sequence repeat polymorphisms. Am J Med Genet 1996;62:262–267.
65 Soo SC, Bain M, Gibson S, White A, Johnstone AP, Nussey S: Isolated congenital ACTH deficiency: a cleavage enzyme defect? Clin Endocrinol 1994;40:555–556.
66 Carey DE: Isolated ACTH deficiency in childhood: lack of response to corticotropin-releasing hormone alone and in combination with arginine vasopressin. J Pediatr 1985;107:925–928.
67 Krude H, Biebermann H, Luck W, Horn R, Brabant G, Gruters A: Severe early-onset obesity, adrenal insufficiency and red hair pigmentation caused by POMC mutations in humans. Nat Genet 1998;19:155–157.
68 Pulichino AM, Vallette-Kasic S, Couture C, Gauthier Y, Brue T, David M, Malpuech G, Deal C, Van Vliet G, De Vroede M, Riepe FG, Partsch CJ, Sippell WG, Berberoglu M, Atasay B, Drouin J: Human and mouse Tpit gene mutations cause early onset pituitary ACTH deficiency. Genes Dev 2003a;17:711–716.

69 Vallette-Kasic S, Brue T, Pulichino AM, Gueydan M, Barlier A, David M, Nicolino M, Malpuech G, Déchelotte P, Deal C, Van Vliet G, De Vroede M, Riepe FG, Partsch CJ, Sippell WG, Berberoglu M, Atasay B, de Zegher F, Beckers D, Kyllo J, Donohoue P, Fassnacht M, Hahner S, Allolio B, Noordam C, Dunkel L, Hero M, Pigeon B, Weill J, Yigit S, Brauner R, Heinrich JJ, Cummings E, Riddell C, Enjalbert A, Drouin J: Congenital isolated adrenocorticotropin deficiency: an underestimated cause of neonatal death, explained by TPIT gene mutations. J Clin Endocrinol Metab 2005;90:1323–1331.

Dr. Jacques Drouin
Laboratoire de Génétique Moléculaire
Institut de Recherches Cliniques de Montréal (IRCM)
110, Avenue des Pins Ouest
Montréal QC H2W 1R7 (Canada)
Tel. +1 514 987 5680, Fax +1 514 987 5575, E-Mail jacques.drouin@ircm.qc.ca

Arzt E, Bronstein M, Guitelman M (eds): Pituitary Today: Molecular, Physiological and Clinical Aspects. Front Horm Res. Basel, Karger, 2006, vol 35, pp 88–101

Long-Term Follow-Up of Prolactinomas: Should Dopamine Agonist Treatment Be Life-Long?

Mirtha Guitelman

Endocrinology Division, Durand Hospital, and Neuroscience's Institute, University of Buenos Aires (INUBA), Buenos Aires, Argentina

Abstract

Dopamine agonists (DA) are the treatment of choice for both micro- and macroadenomas. However, the only limitation of this treatment is the low rate of cure, as well as the long-term therapy usually seen in most patients. In our data, 16% of patients with prolactinomas remained normoprolactinemic after DA withdrawal. The duration of treatment before DA discontinuation was the only statistically significant factor in the cured versus the remission groups. Taking into account all the data from the published series, the rate of cure after DA withdrawal ranged from 5 to 37.5%, and only 1% of tumor regrowth was observed in spite of the high recurrence rate. The board of experts provided recommendations based on the revision of all the data presented at this consensus meeting: dopamine agonists can be routinely withdrawn in patients receiving DA. The length of treatment recommended is 1–3 years, during which DA may be tapered off, and serum PRL levels should be monitored. Tumor disappearance on MRI and the levels of PRL under the lower dose of DA will be determinant for DA withdrawal. Factors influencing prolactinomas outcome are analyzed in the discussion.

The management of prolactinomas with dopamine agonists (DA) was introduced more than 30 years ago [1]. This extremely effective treatment in the control of the hyperprolactinemic syndrome has widely replaced pituitary surgery, and it is nowadays considered the treatment of choice for macro- and microadenomas [2].

Clinical and neuroradiological improvement is often seen in 80% of patients [3].

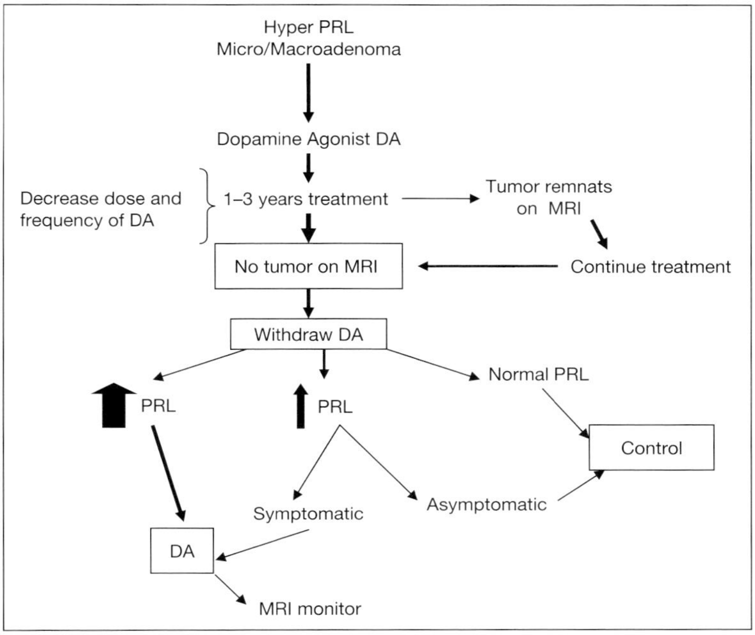

Fig. 1. Algorithm for treated prolactinomas.

Long-term medical treatment of prolactinomas may often result in permanent remission and should not be invariably regarded as a life-long treatment [4]. Although therapeutic withdrawal is recommended, there is no consensus regarding the duration and discontinuation of therapy for treated prolactinomas. We will first show our own data (Endocrinology Division, Durand Hospital, Buenos Aires, Argentina), comparing it with all the data published to date. Then we will provide a discussion about the factors that may be indicators of remission in treated prolactinomas. Finally, we will offer some recommendations together with an algorithm for treated prolactinomas. For this purpose, remarkable neuroendocrinologists (Irene Halperin, MD, Spain; Ruth Mangupli MD, Venezuela; Prof. Ashley Grossman, MD, UK; Marcello D. Bronstein MD, Brazil; Prof. Armando Basso, MD, PhD, Argentina; Prof. Dr. med. Günter K. Stalla, MD, Germany; Marcello Vitale, MD, Argentina) from around the world with an extensive experience in prolactinomas met at 'Pituitary Today' to discuss and reach a consensus on long-term dopamine agonist treatment for prolactinomas.

Data

Our cohort of patients was randomly selected, and records were analyzed retrospectively. We included patients with microadenomas and macroadenomas, some of them operated either prior or after indication of dopamine agonists.

We based the diagnosis of prolactinomas on symptomatic monomeric hyperprolactinemia associated with microadenomas (less than 1 cm on MRI) or macroadenomas (more than 1 cm on MRI).

Serum prolactin (PRL) levels were measured by commercial kits: RIA from 1983 to 1987, immunoradiometric assay (IRMA) from 1987 to 1993, chemiluminiscent (ACS 180 Bayer) assay from 1993 to present date.

Bromocriptine has been used in our country since 1970, and cabergoline was introduced in 1997. Nowadays, 98% of our patients with prolactinomas are being treated with cabergoline as the medical therapy of choice.

Patients with macroprolactinemia and microadenomas were excluded, as screening for macroprolactinemia through the polyethylene glycol method was introduced in our center in 1999.

103 of 157 patients were found to have microadenomas, and 54, macroadenomas. Only 22 patients from this cohort were operated, and 6 of them achieved normal PRL after surgery (27%). The remainder were treated with DA.

100 of 151 (62%) patients withdrew the DA during follow-up – 68 micro- and 32 macroadenomas. Subjects received either bromocriptine BEC (n = 33), cabergoline CAB (n = 19) or both (n = 48). Seven patients were excluded from the analysis because they failed to return to our center, which made a proper follow-up difficult.

Statistical Analysis

Data are presented as means ± SD. Nonparametric statistical methods and Fisher's exact test were applied. Among patients, the analysis was performed with the paired t test. $p < 0.05$ was considered statistically significant.

Results

Considering only microadenomas, 14 patients (n = 62) (22%) remained normo-PRL after DA withdrawal (cured group) and 48 patients showed a rise in PRL levels during follow-up (recurrence group) (table 1).

Age, basal PRL, nadir PRL under treatment, dose of DA, and type of DA showed no significant differences between the two groups. Length of treatment before DA discontinuation was significantly higher in the cured patients compared to patients with recurrence (p = 0.0019). No patients in the normo-PRL group had tumor remnants on MRI; however, in the recurrence group 38% of patients also had a normal MRI. Serum PRL levels during the off-treatment period were significantly lower than basal prolactin (PRL) in the recurrence group.

Table 1. Variables in patients with microadenomas who remained normoprolactinemic or recurred after DA withdrawal

	Cured	Recurrence	p value
n = 62	14 (20%)	48	–
Age/sex	14/F	46/F, 2/M	NS
	mean age: 33.5	mean age: 30.5	
Basal PRL range, ng/ml	123 (70–210)	150 (50–469)	NS
Nadir PRL on DA, ng/ml	14.9 (4.2–30)	18.3 (0.6–71)	NS
Tumor reduction, %	disappearance 100%	no change: 43%	<0.0001
		reduction: 19%	
		disappearance: 38%	
Length of treatment years	mean: 9 years	mean: 4 years	0.0019
Maximum dose of DA	(6–300 months)	(6–240 months)	
	BEC: 7 mg/day	BEC: 5.6 mg/day	NS
	CAB: 0.5 mg/week	CAB: 0.8 mg/week	
Off-DA PRL	mean: 18.9 ng/ml	mean: 107 ng/ml	<0.0001
Period of follow-up, years	2–5	1	–
Time to recurrence, months	–	3–12	–
Pregnancy, %	35.7	33	NS
Menopause, %	21.4	6.2	NS
Surgery	none	3	–

The length of follow-up without treatment was between 2 and 5 years in the cured group. Time to relapse of hyperprolactinemia was between 3 months and 4 years, with a mean of 9 months, in the recurrence group.

Even though a higher incidence of menopause was observed in the cured group, no significant differences were found.

There was a similar incidence of pregnancy in both groups.

As regards the group of macroadenomas, only 1 (3%) of 30 patients remained normo-PRL after DA withdrawal (table 2). Therefore, comparing different variables and obtaining comparative data between the cured and recurrence groups were not possible.

The cured patient was a male who had been first operated and, due to the finding of tumor remnants and high PRL levels after surgery, he had received BEC for 2 years. During that period, the dose was decreased according to the decreasing levels of PRL (<3 ng/ml). The tumor showed features of bleeding and necrosis, and finally disappeared. Serum PRL levels were maintained in low values during a follow-up of 6 years.

In the recurrence group of macroadenomas, significant tumor reduction was observed in 60.7% of cases, 29% developed an empty sella syndrome, and 10.7% had no change.

Table 2. Variables in patients with macroadenomas who remained normoprolactinemic or recurred after DA withdrawal

	Cured	Recurrence
n = 31	1 (3%)	30
Age/sex	male	5/M, 25/F mean age: 32
Basal PRL range, ng/ml	220	Mean PRL: 1,040
Nadir PRL on DA, ng/ml	<3	mean: 31
Tumor reduction, %	tumor disappearance	no change: 10.7% reduction: 60.7% Empty Sella: 29%
Length of treatment, years	2	mean: 4.2 years
Maximum dose of DA	BEC: 5 mg/day	BEC: 10.8 mg/day CAB: 1.1 mg/week
Off-DA PRL	<3 ng/ml	mean: 545 ng/ml
Length of follow-up	6 years	2 months–3 years
Time to recurrence	–	2 months–3 years
Pregnancy, n	–	11
Menopause, n	–	2
Surgery, n	yes	9

No significant differences were observed comparing off-treatment PRL levels to basal ones in the relapsing patients with macroadenoma. Time to relapse was between 2 months and 3 years. Interestingly, 1 patient's PRL relapse occurred only after the third year of follow-up. During the period of normoprolactinemia, she was asymptomatic, and no tumor regrowth was observed. An algorithm for the treatment and diagnosis of prolactiomas is shown in figure 1.

Conclusions

In our cohort of patients with prolactinomas, recurrence after DA withdrawal was observed in 84% of the patients. Normo-PRL after long-term DA treatment was 20% for micro- and 3% for macroadenomas (16% both).

Age, basal PRL levels, nadir PRL under DA, tumor size, dose of DA, type of DA, pregnancy during follow-up, and previous surgery were not good predicting factors of remission.

In patients with microadenomas, the duration of treatment before DA discontinuation was the only statistically significant factor between the cured and the remission groups.

Menopause was more frequent in the group of persistent normo-PRL, but it was not significant enough. No cured patient had tumor remnants, so disappearance of the tumor was a requirement for this group. However, the absence of tumor on MRI does not rule out the chance of recurrence during the period of DA withdrawal.

No tumor reexpansion in spite of the high rate of PRL recurrence was noticed. This could be related to the previous long course of treatment, or it may reflect the low growth rate of prolactinomas and the short length of follow-up without medical therapy.

Published Series on Dopamine Agonist Withdrawal

Table 3 shows the most important data on this topic since 1983 [5–18]. Taking into account all the data from the published series (except for the paper by Colao et al. [17]), the rate of cure after DA withdrawal ranged from 5 to 37.5%, with a mean of 19.4%; the number of patients analyzed was 632; there were 5 series with CAB and 11 with BEC; tumor regrowth was seen in only 2 series, the mean regrowth rate being 1.1%; the mean duration of treatment before withdrawal was 3 years, and the mean period of withdrawal was 1.7 years.

Colao et al. [17] found the highest cure rate: 67% for microadenomas and 57% for macroadenomas in a cohort of 200 hyperprolactinemic patients treated with cabergoline for a mean period of 2 years and a follow-up without treatment of between 2 and 5 years. In this study protocol described by Colao et al. [17], withdrawal of long-term cabergoline therapy for prolactinomas had higher rates of normo-PRL than those reported in the literature. They justified these differences by the lack of criteria for timing withdrawal, the systematic evaluation of discontinuation outcome by prolactin suppression and tumor shrinkage, and the effective antitumoral effect of cabergoline. However, they excluded pregnant, previously surgically treated and radiation therapy receiving patients, conditions which may influence the course of prolactinomas. Patients with macroprolactinemia were not excluded, however, and it is a common condition observed in patients with hyperprolactinemia.

In the study of Colao et al. [17], the initial population of patients analyzed was 354, and the final number was 200, due to the exclusion of patients which became pregnant, those whose prolactin failed to reach normal levels, and those without significant shrinkage of the tumor.

The high rate of cured cases is true for a selected group of patients with prolactinomas, that is, those with normal PRL levels and MRI during treatment with cabergoline. This rate would have been lower if they had included all the population analyzed since the beginning, in which case the rate of cure would have been 36% for macro and microadenomas.

Table 3. Published series on DA withdrawal

Reference	n	Normo-PRL %	Regrowth %	Length of treatment	Withdrawal	Type of tumor	Agent
Zarate et al. [5], 1983	16	37.5	0	2 years	4 years	macro/micro	BEC
Johnston et al. [6], 1984	15	7	7	3.7 years	1 year	macro/micro	BEC
Ho et al. [7], 1985	15	26	0	5 years	¿	micro	BEC
Moriondo et al. [8], 1985	36	11	0	1 year	14–30 months	micro	BEC
	18	22		2 years	8–28 months		
Wang et al. [9], 1987	24	24	0	2 years	1–4 years	micro/macro	BEC
Rasmussen et al. [10], 1987	75	5	0	24–65 months × 24 months	>6 months	micro/macro	BEC
van't Veerlat [11], 1991	12	25 8	7	3.5–7 years	1 year >1 year	macro	BEC
Ferrari et al. [12], 1992	32	31	0	× 14 months (3–52 months)	1 year	micro/macro	CAB
Muratori et al. [13], 1997	26	10	0	1–2 years	38–60 months	micro	CAB
Canavo et al. [14], 1999	27	18	0	2 years	1 year	macro/micro	CAB
Karunakaran et al. [15], 2001	116	15	0	7.5 years	>3 months	micro	BEC
Passos et al. [16], 2002	131	20	0	4 years	3.6 years	macro/micro	BEC
Colao et al. [17], 2003	175	57 macro 67 micro	0	× 48 months	2–5 years	macro micro	CAB
Biswas et al. [18], 2005	89	31	0	× 37 months	>12 months	micro	BEC 22 CAB 67

Macro = Macroadenoma; micro = microadenoma.

Discussion

Studies of DA therapy suggested that normoprolactinemic remission was possible following discontinuation of therapy in prolactinomas (7–37.5%). Therefore, the controversy as to whether DA therapy should be life-long remained.

According to a survey on DA withdrawal policy in the UK, 80% of endocrinologists routinely withdraw DA when treating microprolactinomas [19]. The criteria as to why DA was withdrawn were resolution of tumor on imaging, normal PRL after at least 3 years of treatment, normal PRL maintained on the smallest dose of DA, menopause, and pregnancy. Regarding macroadenomas, only 16% withdraw therapy following the same criteria as for microadenomas.

There are some factors that influence the course, evolution, and resolution of prolactinomas: antitumoral effects of DA, natural history of microadenomas, length of DA therapy, pregnancy, menopause, serum basal PRL levels, and nadir PRL levels under DA treatment, tumor shrinkage and disappearance on MRI during treatment, previous surgery, and type of drug used.

We analyze each of these factors and discuss their true influence in terms of prolactinoma remission:

Concerning the antitumoral effects of DA, in 1980 Maurer et al. [20] demonstrated that bromocriptine causes an inhibition of PRL mRNA transcription and PRL synthesis in normal and tumor cells; Basetti et al. [21] and Tindall et al. [22] suggested that short-term therapy with bromocriptine is associated with a reversible reduction of organelles involved in PRL synthesis. Later, Gen et al. [23] showed that long-term bromocriptine treatment causes tumor atrophy, necrosis, fibrosis, hyalinosis, and inflammatory cell infiltration, suggesting cytocidal effects. Clearly, DA inhibits PRL secretion and cell proliferation, but whether or not the antitumoral action of DA is connected with the induction of apoptosis is controversial [24, 25].

Pituitary apoplexy is a rare life-threatening condition caused by a sudden infarction or hemorrhagic necrosis of the pituitary containing an adenoma [26]. A wide variety of conditions can trigger apoplexy, such as pituitary irradiation, general anesthesia, traumatic head injury, pituitary stimulatory tests and a wide variety of medications including DA [27].

The fact that the duration of DA therapy may influence the outcome of prolactinomas regarding the results of short- and long-term therapy in morphological studies is easy to understand.

Tumor regrowth was observed after short-term BEC therapy [28, 29]. Conversely, long-term DA therapy is associated with the absence of tumor reexpansion [5–18]. Duration of therapy may be an important determinant of remission, as we demonstrate in our cohort of patients with microadenomas; however, there is still controversy concerning the exact duration of DA therapy.

In all the published data the mean duration of therapy was 3 years. The disappearance of the tumor on MRI is an important end point when deciding upon treatment discontinuation, so duration of treatment will vary depending on imaging findings during follow-up.

A factor that may influence remission of prolactinomas is their natural history. This conclusion is a result of several series where patients with microprolactinomas were observed for a long period of time without DA [30–36].

It is thought that pituitary apoplexy could be involved in such spontaneous tumor resolution. Jeffcoate et al. [37] found that 33% of women with microadenomas followed for 15 years without any treatment had spontaneous normalization of PRL levels.

As for macroadenomas, spontaneous regression is less frequent, but there are published case reports of pituitary apoplexy as presentation of macroprolactinomas [26].

Pregnancy is another condition that can change the resolution of treated prolactinomas. The pituitary gland grows more than a hundred percent during pregnancy due to high levels of estrogen [38], and some patients, especially those with macroadenomas with no treatment before pregnancy, are more likely to have neuro-ophthalmological complications [39, 40].

In contrast with these phenomena, there are several arguments supporting the idea that pregnancy may favorably influence the outcome of prolactinomas [41, 42]. Several studies showed that the PRL level decreases or even normalizes after pregnancies in women with prolactinomas [43]. Jeffcoate et al. [37] found a higher rate of remission in women with microadenomas who became pregnant during a follow-up of 15 years (35 vs. 14%).

Karunakaran et al. [15] suggested that females who become pregnant may have a higher chance of normalizing their PRL levels (33 vs. 18%).

Regarding tumor changes related to pregnancy, 27% (n = 41) and 34% (n = 44) of patients showed some degree of reduction after delivery.

Whether the effects of pregnancy are a result of a mechanical process or they are due to estrogen per se has not been established yet. It has recently been demonstrated that 17ß-estradiol upregulates the expression of the Fas/FasL system in anterior pituitary cells and increases Fas-induced apoptosis in lactotropes [45]. However, enough data are lacking to determine any statistically significant influence of pregnancy concerning maintenance of normoprolactinemia after drug withdrawal.

Another factor that has been implicated in remission is menopause. It is reasonable to assume that lower levels of estrogen during menopause may have a positive effect in terms of tumor growth and response to treatment. Karunakaran et al. [15] showed a significant chance of normalizing PRL levels in menopausal patients with microadenomas. In spite of the fact that the number of menopausal patients was higher in our group of cured patients with microadenomas, it was not enough to achieve a significant statistical value. An interesting observation is the fact that patients with prolactinomas who reach menopause had been under treatment longer before DA withdrawal, and this might explain the high tendency of cure in this group.

Pretreatment PRL levels were not good remission factors of cure in our cohort of patients with microadenomas. Biswas et al. [18] found a significant risk of relapse among patients with higher levels of PRL before treatment, regardless of the size of the lesion. The relationship between baseline PRL levels and risk of relapse has also been observed in some of the previous studies [8, 17], but not in others [16].

Nadir PRL during DA treatment was significantly lower in the group of patients with prolactinomas who became normoprolactinemic after DA withdrawal in the series of Colao et al. [17]. These results have not been repeated by us or others, where nadir PRL seemed to be a poor parameter of remission in treated prolactinomas [16–18].

Tumor shrinkage and pituitary tumor disappearance on MRI were the best predicting factors of remission. In our group of patients who remained normoprolactinemic MRI was normal without tumor remnants. However, 38% of micro- and 29% of macroadenomas had no further tumor on imaging studies, in spite of the recurrence of hyperprolactinemia when DA was withdrawn. These findings suggest that tumor remnants, smaller than the size the highest MRI resolution would show, are still present. It should be noted that microscopic bone and dural invasion is a common finding in surgical studies on prolactinomas [46, 47].

Regarding pituitary imaging studies, it is interesting to note that several published studies used CT for both diagnosis and follow-up of prolactinomas before and after DA therapy [5, 6, 10, 13]. MRI is superior to CT in diagnosing pituitary microadenomas, in the follow-up of medically treated prolactinomas and in the study of extrasellar involvement of pituitary adenomas [48–50, 54]. Nevertheless, despite its superiority over CT, MRI has a high incidence of microincidentalomas (10–15%), which may result in over-diagnosis of microadenomas in the healthy population [51–55].

Previous surgery is a fact not always discussed as positively influencing treated prolactinomas. However, there are no available data confirming this view.

Bromocriptine vs. cabergoline: is there any difference between patients who receive one or the other drug in terms of persisted normoprolactinemia after withdrawal?

Colao et al. [17] attributed the high rate of stable normoprolactinemia without tumor recurrence to an effective antitumoral effect of cabergoline. They also demonstrated that macroprolactinoma shrinkage during cabergoline treatment is greater in naïve patients than in patients pretreated with other DA [56]. In our own cohort of patients, we cannot demonstrate any superiority for either bromocriptine or cabergoline in terms of cure after long-term treatment.

However, it is true that cabergoline leads to a longer period of remission once the drug has been withdrawn [14]. This is related to the longer half-life of cabergoline compared to BEC and its longer period of inhibition on PRL levels compared to bromocriptine [57, 58].

The length of follow-up during DA withdrawal was variable in all the series published, with a mean of 4 years (table 3). However, a recommended exact monitoring period was not established. In our group of cured patients the maximum length of follow-up was 5 years; it should be noted that 1 patient with micro- and another with macroadenoma recurred after 4 and 3 years, respectively. Slow growth of prolactinomas has been recognized, therefore increasing PRL levels even in patients without symptoms may anticipate recurrence.

The Board of Experts reaches some conclusions and recommendations:

(1) Dopamine agonists should be tapered off rather than abruptly discontinued.
(2) No evidence of tumor on MRI seems to be the most reliable criteria for DA withdrawal.
(3) The length of treatment before DA discontinuation might be established in the range of 1–3 years.
(4) Patients with a sustained elevated PRL in spite of clinical recovery should be studied for PRL isoforms (macroprolactin) before increasing the DA dose.
(5) Pregnancy and menopause are periods where patients could be studied without treatment.

Acknowledgements

Natalia Garcia Basavilbaso, a fellowship at the Division of Endocrinology of the Hospital Durand, helped with the revision of all the records and the statistical analysis.

Board of Experts

Prof. Armando Basso, MD, PhD: Director of Neuroscience's Institute, University of Buenos Aires. President of the WFNS Foundation. Marcello D. Bronstein: Division of Endocrinology and Metabolism Hospital das Clinicas, University of Sao Paulo Medical School. Prof. Ashley Grossman: President of the European Neuroendocrine Association. Department of Endocrinology Barts and the London Queen Mary University of London.

Prof. Dr. med. Günter K. Stalla, MD: Department of Internal Medicine/Endocrinology and Clinical Chemistry Munich, Germany.

Irene Halperin, MD: Endocrinology and Nutrition Unit of the Hospital Clínic, Barcelona, Spain.

Ruth Mangupli, MD: Neurosurgery Unit from the University Hospital of Caracas, Venezuela.

Marcello Vitale, MD: Neuroendocrinology Area at the Division of Neurosurgery, Hospital Santa Lucía, Buenos Aires.

References

1 Besser GM, Parke L, Edwards CRW, Forsyth LA, Mc Neilly AS: Galactorrhea: successful treatment with reduction of plasma prolactin levels by brom-ergocryptine. J Clin Endocrinol Metabol 1984;58:268–273.

2 Spark RF, Baker R, Bienfang DC, Bergland R: Bromocriptine reduces pituitary tumor size and hypersecretion: requiem for pituitary surgery? J Am Med Assoc 1982;247:311–316.

3 Molitch ME: Medical treatment of prolactinomas. Endocrinol Metab Clin North Am 1999;28: 143–169

4 Faglia G: Should dopamine agonist treatment for prolactinomas be life-long? Clin Endocrinol (Oxf) 1991;34:173–174.

5 Zarate A, Canales ES, Cano C, Pilonieta CJ: Follow-up of patients with prolactinomas after discontinuation of long-term therapy with bromocriptine. Acta Endocrinol (Copenh) 1983;104:139–142.

6 Johnston DG, Hall K, Kendall-Taylor P, Patrick D, Watson M, Cook DB: Effect of dopamine agonist withdrawal after long-term therapy in prolactinomas: studies with high-definition computerized tomography. Lancet 1984;ii:187–192.

7 Ho KY, Smythe GA, Compton PJ, Lazarus SL: Long-term bromocriptine therapy may restore the inhibition control of prolactin release in some patients with pathological hyperprolactinemia. Aust NZ J Med 1985;15:213–219.

8 Moriondo P, Travaglini P, Nissim M, Conti A, Faglia G: Bromocriptine treatment of microprolactinomas: evidence of stable prolactin decrease after drug withdrawal. J Clin Endocrinol Metab 1985;60:764–772.

9 Wang C, Lam KS, Ma JT, Chan T, Liu MY, Yeung RT: Long-term treatment of hyperprolactinaemia with bromocriptine: effect of drug withdrawal. Clin Endocrinol (Oxf) 1987;27:363–371.

10 Rasmussen C, Bergh T, Wide L: Prolactin secretion and menstrual function after long-term bromocriptine treatment. Fertil Steril 1987;48:550–554.

11 van't Verlaat JW, Croughs RJ: Withdrawal of bromocriptine after long-term therapy for macroprolactinomas, effect on plasma prolactin and tumour size. Clin Endocrinol (Oxf) 1991;34:175–178.

12 Ferrari C, Paracchi A, Mattei AM, de Vincentiis S, Dálberton A, Crosignani P: Cabergoline in the long-term therapy of hyperprolactinemia disorders. Acta Endocrinol (Copenh) 1992;126:489–494.

13 Muratori M, Arosio M, Gambino G, Romano C, Biella O, Faglia G: Use of Cabergoline in the long-term treatment of hyperprolactinemic and acromegalic patients. J Endocrinol Invest 1997;20: 537–546.

14 Cannavo S, Curto L, Squadrito S, Almoto B, Vieni A, Trimarchi F: Cabergoline: a first choice treatment in patients with previously untreated prolactin-secreting pituitary adenomas. J Endocrinol Invest 1999;22:354–359.

15 Karunakaran S, Page RC, Wass JA: The effect of the menopause on prolactin levels in patients with hyperprolactinaemia. Clin Endocrinol (Oxf) 2001;54:295–300.

16 Passos VQ, Souza J, Musolino N, Bronstein MD: Long term follow-up of prolactinomas: normoprolactinemia after bromocriptine withdrawal. J Clin Endocrinol Metab 2002;87:3578–3582.

17 Colao A, Di Sarno A, Cappabianca P, Di Somma C, Pivonello R, Lombardi G: Withdrawal of long-term cabergoline therapy for tumoral and nontumoral hyperprolactinemia. N Engl J Med 2003;349:2023–2033.

18 Biswas M, Smith J, Jadon D, Mc Ewan P, Rees DA, Evans LM, Scanion MF, Davies JS: Long-term remission following withdrawal of dopamine agonist therapy in subjects with microprolactinomas. Clin Endocrinol (Oxf) 2005;63:26–31.

19 Mehmet S, Powrie JK: A survey of dopamine agonist withdrawal policy in UK endocrinologist treating patients with prolactinomas. Clin Endocrinol (Oxf) 2003;58:111–113.

20 Maurer RA: Dopaminergic inhibition of prolactin synthesis and prolactin messenger RNA accumulation in cultured pituitary cells. J Biol Chem 1980;255:8092–8097.

21 Bassetti M, Spada A, Pezzo G, Giannattasio G: Bromocriptine treatment reduces the cell size in human macroprolactinomas: a morphometric study. J Clin Endocrinol Metab 1984;58: 268–273.
22 Tindall GT, Kovacs K, Horvath E, Thorner MO: Human prolactin-producing adenomas and bromocriptine: a histological, immunocytochemical, ultrastructural, and morphometric study. J Clin Endocrinol Metab 1982;55:1178–1183.
23 Gen M, Uozumi T, Ohta M, Ito A, Kajiwara H, Mori S: Necrotic changes in prolactinomas after long-term administration of bromocriptine. J Clin Endocrinol Metab 1984;59:463–470.
24 Stefaneanu L, Kovacs K, Scheithauer BW, Kontogeorgos G, Riehle DL, Sebo TJ, Murray D, Vidal S, Tran A, Buchfelder M, Fahlbusch R: Effect of dopamine agonists on lactotroph adenomas of the human pituitary. Endocr Pathol 2000;11:341–352.
25 Gruszka A, Pawlikowski M, Kunert-Radek J: Anti-tumoral action of octreotide and bromocriptine on the experimental rat prolactinoma: antiproliferative and pro-apoptotic effects. Neuroendocrinol Lett 2001;22:343–348.
26 Lubina D, Olchovsky M, Berezin Z, Ram M, Hadani M, Shimon I: Management of pituitary apoplexy: clinical experience with 40 patients. Acta Neurochir 2005;147:151–157.
27 Knoepfelmacher M, Gomes MC, Melo ME, Mendonça BB: Pituitary apoplexy during therapy with cabergoline in an adolescent male with prolactin-secreting macroadenoma Pituitary 2005;7: 83–87.
28 Thorner MO, Perryman RL, Rogol AD, Conway BP, Macleod RM, Login IS, Morris JL: Rapid changes of prolactinoma volume after withdrawal and reinstitution of bromocriptine. J Clin Endocrinol Metab 1981;53:480–483.
29 Orrego JJ, Chandler WF, Barkan AL: Rapid re-expansion of a macroprolactinoma after early discontinuation of bromocriptine. Pituitary 2000;3:189–192.
30 Koppelman MC, Jaffe MJ, Rieth KG, Caruso RC, Louriaux DL: Hyperprolactinemia, amenorrhea, and galactorrhea: a retrospective assessment of twenty-five cases. Ann Intern Med 1984;100: 115–121.
31 March CM, Kletzky OA, Davajan V, Teal J, Weiss M, Apuzzo ML, Marrs RP, Mishell DR Jr: Longitudinal evaluation of patients with untreated prolactin-secreting pituitary adenomas. Am J Obstet Gynecol 1981;139:835–844.
32 Schlechte J, Dolan K, Sherman B, Chapler F, Luciano A: The natural history of untreated hyperprolactinemia: a prospective analysis. J Clin Endocrinol Metab 1989;68:412–418.
33 Sisam DA, Sheehan JP, Sheeler LR: The natural history of untreated microprolactinomas. Fertil Steril 1987;48:67–71.
34 Rjosk HK, Fahlbusch R, von Werder K: Spontaneous development of hyperprolactinemia. Acta Endocrinol (Copenh) 1982;100:333–336.
35 Martin TL, Kim M, Malarkey WB: The natural history of idiopathic hyperprolactinemia. J Clin Endocrinol Metab 1985;60:855–858.
36 Weiss MH, Teal J, Gott P, Wycoff R, Yadley R, Apuzzo ML, Giannotta SL, Kletzky O, March C: Natural history of microprolactinomas: six-year follow-up. Neurosurgery 1983;12:180–183.
37 Jeffcoate WJ, Pound N, Sturrock ND, Lambourne J: Long-term follow-up of patients with hyperprolactinaemia. Clin Endocrinol (Oxf) 1996;45:299–303.
38 Villarreal JZ: Pituitary gland growth during normal pregnancy: an in vivo study using magnetic resonance imaging. Am J Med 1988;85:217–220.
39 Molitch ME: Pregnancy and the hyperprolactinemic woman. N Engl J Med 1985;312:1365–1370.
40 Molitch ME: Management of prolactinomas during pregnancy. J Reprod Med 1999;44: 1121–1126.
41 Badawy SZ, Marziale JC, Rosenbaum AE, Chang JK, Joy SE: The long-term effects of pregnancy and bromocriptine treatment on prolactinomas – the value of radiologic studies. Early Pregnancy 1997;3:306–311.
42 Ahmed M, al-Dossary E, Woodhouse NJ: Macroprolactinomas with supraselar extension: effect of bromocriptine withdrawal during one or more pregnancies. Fertil Steril 1992;58:492–497.
43 Crosignani PG, Mattei AM, Severini V, Cavioni V, Maggioni P, Testa G: Long-term effects of time, medical treatment and pregnancy in 176 hyperprolactinemic women. Eur J Obstet Gynecol Reprod Biol 1992;44:175–180.

44 Bronstein MD, Salgado LR, Musolino NR: Medical management of pituitary adenomas: the special case of management of the pregnant woman. Pituitary 2002;5:99–107.
45 Jaita AG, Candolfi M, Zaldivar V, Zárate S, Ferrari L, Pisera D, Castro MG, Seilicovich A: Estrogens up-regulate the Fas/FasL apoptotic pathway in lactotropes. Endocrinology 2005;146: 4737–4744.
46 Selman WR, Laws ER Jr, Scheithauer BW, Carpenter SM: The occurrence of dural invasion in pituitary adenomas. Neurosurgery 1986;64:402–407.
47 Meij BP, Lopes MB, Ellegala DB, Alden TD, Laws ER: The long-term significance of microscopic dural invasion in 354 patients with pituitary adenomas treated with transsphenoidal surgery. J Neurosurgery 2002;96:195–208.
48 Juliani G, Avataneo T, Potenzoni F, Sorrentino T: CT and MRI compared in the study of hypophysis. Radiol Med (Torino) 1989;77:51–64.
49 Guy RL, Benn JJ, Ayers AB, Bingham JB, Lowy C, Cox TC, Sonksen PH: A comparison of CT and MRI in the assessment of the pituitary and parasellar region. Clin Radiol 1991;43:156–161.
50 Webb SM, Ruscalleda J, Schwarzstein D, Calaf-Alsina J, Rovira A, Matos G, Puig-Domingo M, de Leiva A: Computerized tomography versus magnetic resonance imaging: a comparative study in hypothalamic-pituitary and parasellar pathology. Clin Endocrinol (Oxf) 1992;36:459–465.
51 Teramoto A, Hirakawa K, Sanno N, Osamura Y: Incidental pituitary lesions in 1,000 unselected autopsy specimens. Radiology 1994;193:161–164.
52 Reincke M, Allolio B, Saeger W, Menzel J, Winkelmann W: The 'incidentaloma' of the pituitary gland. Is neurosurgery required? JAMA 1990;263:2772–2776.
53 Aron DC, Howlett TA: Pituitary incidentalomas. Endocrinol Metab Clin North Am 2000;29: 205–221.
54 Stein AL, Levenick MN, Kletzky OA: Computed tomography versus magnetic resonance imaging for the evaluation of suspected pituitary adenomas. Obstet Gynecol 1989;73:996–999.
55 Chanson P, Daujat F, Young J, Bellucci A, Kujas M, Doyon D, Schaison G: Normal pituitary hypertrophy as a frequent cause of pituitary incidentaloma: a follow-up study. J Clin Endocrinol Metabol 2001;86:3009–3015.
56 Colao A, Di Sarno A, Landi ML, Scavuzzo F, Cappabianca P, Pivonello R, Volpe R, Di Salle F, Cirillo S, Annunziato L, Lombardi G: Macroprolactinoma shrinkage during cabergoline treatment is greater in naïve patients than in patients pretreated with other dopamine agonists: a prospective study in 110 patients. J Clin Endocrinol Metabol 2000;85:2247–2252.
57 Webster J, Piscitelli G, Polli A, Ferrari CI, Ismail I, Scanlon MF: A comparison of cabergoline and bromocriptine in the treatment of hyperprolactinemic amenorrhea. Cabergoline Comparative Study Group. N Engl J Med 1994;331:904–909.
58 Verhelst J, Abs R, Maiter D, van den Bruel A, Vandeweghe M, Velkeniers B, Mockel J, Lamberigts G, Petrossians P, Coremans P, Mahler C, Stevenaert A, Verlooy J, Raftopoulos C, Beckers A: Cabergoline in the treatment of hyperprolactinemia: a study in 455 patients. J Clin Endocrinol Metab 1999;84:2518–2522.

Dr. Mirtha Guitelman
División Endocrinología, Hospital Carlos G. Durand
Av. Díaz Vélez 5044
Buenos Aires 1405 (Argentina)
Tel. +54 11 4982 5212, Fax +54 11 4958 4377, E-Mail mguitelman@speedy.com.ar

Arzt E, Bronstein M, Guitelman M (eds): Pituitary Today: Molecular, Physiological and Clinical Aspects. Front Horm Res. Basel, Karger, 2006, vol 35, pp 102–114

Ghrelin: From Somatotrope Secretion to New Perspectives in the Regulation of Peripheral Metabolic Functions

F. Broglio[a], *F. Prodam*[a], *F. Riganti*[a], *G. Muccioli*[b], *E. Ghigo*[a]

[a]Division of Endocrinology and Metabolism, Department of Internal Medicine, [b]Department of Anatomy, Pharmacology and Forensic Medicine, University of Turin, Turin, Italy

Abstract

Ghrelin, a peptide predominantly produced by the stomach, has been discovered as natural ligand of the GH secretagogue (GHS) receptor type 1a (GHS-R1a), suggesting the existence a new endogenous modulator of somatotrope secretion. Subsequently, ghrelin turned out to exert pleiotropic actions, consistent with the widespread distribution of ghrelin and GHS-R expression in central and peripheral tissues. Despite that the binding to GHS-R1a requires ghrelin to be acylated in serine 3, some ghrelin actions are independent of such acylation; thus suggesting the possibility of the existence of other GHS-R subtypes. Ghrelin secretion (70% in its unacylated form) is mainly under metabolic control being modulated by glucose, insulin and feeding. On the other hand, ghrelin influences energy metabolism acting both as a central orexigenic factor and directly on the endocrine pancreas, liver and adipose tissue. Recently, another gastric hormone derived from the same ghrelin gene has been isolated and named obestatin. Obestatin in rats resulted in reduced food intake, jejunal contraction and body weight gain, via specific distinct receptors. Thus, all these data indicate that we are exploring a very complex system deeply involved in the modulation of metabolic functions, whose understanding will probably increase our knowledge about diabetes mellitus and the metabolic syndrome.

Ghrelin is a 28-aminoacid peptide predominantly produced by the stomach that has been discovered as a natural ligand of the orphan GH secretagogue (GHS) receptor (GHS-R) type 1a and is therefore supposed to play a major role in the control of somatotrope function, although able to play also some other neuroendocrine actions such as stimulation of lactotrope and corticotrope secretion [1, 2]. As these ghrelin actions, mediated by the binding and activation of the GHSR-1a, have been shown to depend from its unique acylation in serine 3 [3], it has been initially stated that acylated ghrelin is the only biologically active hormonal form [1, 2]. On the contrary, instead, recent data have clearly revealed a different picture: ghrelin mostly circulates in an unacylated form that despite being unable to bind the GHS-R1a, is, as a matter of fact, a biologically active molecule, although devoid of neuroendocrine actions [1, 2]. This evidence is consistent with the existence of GHS-R subtypes that are activated by ghrelin independent of its acylation [1, 2].

GHS-Rs are expressed in the hypothalamus-pituitary unit but also in other areas of the central nervous system and in peripheral tissues, particularly in the endocrine pancreas [1, 2]; this GHS-R distribution explains the pleiotropic actions of ghrelin. Ghrelin has been shown to exert a potent orexigenic activity that, in functional antagonism to leptin, likely signals the metabolic balance and manages the neuroendocrine and metabolic response to starvation [4]. Evidence that acylated ghrelin stimulates appetite and food intake and centrally modulates energy expenditure attracted enormous interest in science; the obvious dream is that ghrelin analogues acting as agonists or antagonists would be useful for the treatment of anorexia/cachexia and obesity, respectively [1, 2, 5]. This enthusiasm has, however, been dampened by reports showing that ghrelin or GHS-R1a KO mice are neither anorectic nor dwarfs [6, 7]. Other studies, however, found that ghrelin deletion impacts on energy balance and, in the long term, GHS-R1a KO mice show alterations such as impaired ability to maintain euglycemia and ventricular dysfunction [6, 8].

The central action of ghrelin in the modulation of appetite, food intake and energy expenditure per se indicates a relevant impact of ghrelin on metabolism. However, it has also been demonstrated that ghrelin, either acylated or not, also exerts direct peripheral metabolic actions, namely on the endocrine pancreas, liver and adipose tissue [1, 2]. These actions are those now receiving particular attention for their obviously important clinical implications.

Control of Ghrelin Secretion

The stomach is the major source of circulating ghrelin levels as shown by their clear decrease after gastrectomy [9]. In humans, ghrelin secretion undergoes marked variations throughout the day with remarkable transient reductions after meals [1, 2, 10]. If ghrelin secretion is age- and/or gender-dependent is still a matter of debate [1, 2].

Circulating total ghrelin levels are negatively associated with body mass index; ghrelin secretion is increased in anorexia and cachexia, reduced in obesity and normalized by recovery of ideal body weight [1, 2]. An exception to the negative association between body mass and ghrelin secretion is Prader-Willi syndrome where obesity is surprisingly associated with ghrelin hypersecretion [11].

Ghrelin secretion is decreased by either intravenous or oral glucose load as well as during euglycemic hyperinsulinemic clamp and even after insulin-induced hypoglycemia [1, 2, 4]. The inhibitory influence of overexposure to insulin on ghrelin secretion is consistent with the strong negative association between ghrelin and insulin levels that had been predicted by the negative correlation between ghrelin levels and body mass index [12]; whether insulin and glucose per se play a direct inhibitory role on ghrelin secretion is, however, still unclear as is the influence, if any, of amino acids and lipids [1, 2, 4, 13].

The activation of somatostatin receptors is probably the most important inhibitory secretory input on ghrelin that, in turn, stimulates somatostatin release at both the hypothalamic and pancreatic levels, thus showing a feedback link between these two peptides [14–16]. Ghrelin secretion is on the other hand under stimulatory control by the cholinergic system [17]. The list of other factors that seem to modulate ghrelin expression and/or secretion includes GLP-1, PYY, oxyntomodulin, urocortin, thyroid hormones and gonadal steroids, namely testosterone [1, 2, 18]. The role of glucocorticoids, if any, is still unclear while leptin has been shown not to affect ghrelin levels [19–23].

Ghrelin hyposecretory states (e.g. after gastrectomy or gastric bypass, obesity, diabetes mellitus, hyperthyroidism) and hypersecretory states (anorexia, cachexia, malnutrition, Prader-Willi syndrome) including ghrelinomas (mostly gastro-intestinal carcinoids) have been described but the clinical consequences of the altered ghrelin secretion remain to be demonstrated [1, 2, 24].

Ghrelin and GHS Receptor(s) Distribution

As stated above, the stomach is the major source of circulating ghrelin levels, and, within the stomach ghrelin expression has been localized in the X/A-like cells that account for 20–25% of all the endocrine cells in the oxyntic mucosa [1, 2]. Ghrelin expression has been demonstrated in several other tissues, in particular in the endocrine pancreas that has been demonstrated to be the major source of ghrelin in fetal life [25]. Indeed, ghrelin-secreting cells represent a new physiological population within the endocrine pancreatic islets and have been named ‘epsilon cells’ [1, 2].

The GHS-R 1a is encoded by a single gene found at chromosomal location 3q26.2 [26]. Two types of GHS receptor complementary DNAs (cDNA) as a result of alternate processing of a pre-mRNA have been identified and named receptor 1a and receptor 1b [27]. The GHS1b receptor is not activated by GHS and its functional role remains unclear [27].

As anticipated, the acyl group at serine 3 (more frequently an octanoyl group) of the ghrelin molecule is essential for binding to activating the GHS-R1a [28]. Interestingly, GHS-R1a is also bound by adenosine (that weakly activates it) and cortistatin, a natural analogue of somatostatin that, instead, does not recognize the GHS-R1a [29].

The GHS-R1a is remarkably expressed in the hypothalamus-pituitary unit consistent with the potent influence on the anterior pituitary function as well as on the control of appetite, food intake and energy balance [1, 2]. Besides its distribution in other central areas, GHS-R1a expression and binding sites have also been demonstrated in several peripheral tissues such as the stomach, intestine, pancreas, thyroid, gonads, adrenal, kidney, heart and vasculature as well as in many endocrine and nonendocrine tumors and cell lines [1, 2].

An increasing amount of data strongly supports the existence of GHS receptor subtypes. Within the cardiovascular system as well as in the pancreas and adipose tissue radiolabelled acylated ghrelin binding is inhibited by unacylated as well as acylated ghrelin [1, 2]. As unacylated ghrelin exerts some metabolic, cardiovascular and cellular actions [4, 30], this evidence strongly supports the existence of a ghrelin receptor subtype activated by the peptide independently of its acylation.

On the other hand, in rat and human cardiovascular system and probably in other endocrine and nonendocrine tissues, the existence of binding sites able to bind only peptidyl GHS has been demonstrated. These binding sites are not ghrelin receptors being unable to bind ghrelin and the non-peptidyl GHS MK-0677 [2, 31]. The amino acid sequence of this receptor is that of CD36, a multifunctional class B scavenger receptor also known as glycoprotein IV that it has been implicated in multiple physiological functions (e.g. cellular adhesion, fatty acid and lipid transportation and antigen presentation) and pathological processes related to macrophage foam cell formation and the pathogenesis of atherosclerosis [31].

Ghrelin, the Endocrine Pancreas and Glucose Metabolism

GHS-R1a, 1b and also ghrelin receptor subtypes able to bind both acylated and unacylated ghrelin have been demonstrated to be expressed within the human pancreas [2].

On the other hand, the expression of ghrelin itself has been shown within the endocrine pancreas, variably localized in α, β and non-β, non-α pancreatic cells [1, 2, 32]. Ghrelin expression at the pancreatic level anticipates that in the stomach, in fact, the pancreas seems to be the major source of circulating ghrelin levels during fetal life [25]. More recently, a group of ghrelin-secreting cells within the pancreatic islets and named epsilon cells has been proposed as a new physiological cell population [1, 2, 33, 34]. Islet insulin and ghrelin cells have been shown to share a common progenitor and Nkx2.2 and Pax4 seem to be required to specify or maintain differentiation of the β cell fate [35]. Thus, there would be a genetic component underlying the balance between insulin and ghrelin cell populations [32, 35].

Given the picture of ghrelin and GHS-R expression within the endocrine pancreas, it is likely to hypothesize that ghrelin system play a role, at least via autocrine/paracrine actions, in the control of the endocrine pancreatic function [32].

In fact, it had been demonstrated that synthetic GHSs have a diabetogenic effect in obese Zucker rats and the chronic treatment with MK-0677, a nonpeptidyl GHS, induces insulin resistance and hyperglycemia in healthy elderly subjects [1, 2]. Subsequently, it has been reported that ghrelin is able to counteract the inhibitory effect of insulin on gluconeogenesis in a hepatoma cell line [36]. Some studies reported that ghrelin is able to stimulate insulin secretion from isolated rat pancreatic islets and in rats in vivo [1, 2]. However, many in vitro, ex vivo and in vivo studies reported that acylated ghrelin inhibits insulin secretion in rats [37]. Particularly, it has been demonstrated that acylated ghrelin inhibits both basal and glucose-stimulated insulin secretion and intracellular calcium from rat islets [38]. Moreover, acylated ghrelin dose dependently attenuates the insulin response to glucose load in rats in vivo [1, 2]; similarly, it inhibits insulin secretion from isolated rat pancreas perfused in situ after stimulation with glucose, arginine, and carbachol [39]. On the other hand, counteraction of endogenous ghrelin by GHS-R antagonists or specific antiserum enhances the insulin response to glucose load, in the meantime reducing the elevation of glucose levels in rats in vivo [38].

Consistent with data in animals, in humans the acute administration of acylated ghrelin is followed by a significant, although transient, reduction in insulin levels while glucagon secretion is unaffected [1, 2]. Pretreatment with ghrelin has also been found to be able to reduce the insulin response to arginine but not that to oral glucose load [2].

In these experimental conditions, acylated ghrelin administration has been shown to increase plasma glucose levels even before the transient insulin reduction [1, 2] and to amplify the hyperglycemic effect of arginine [2, 40].

As the hyperglycemic effect of acylated ghrelin is not coupled with stimulation of glucagon release, it has been hypothesized that ghrelin activates glycogenolysis either indirectly via stimulation of catecholamine release or by acting directly on the hepatocytes. In fact, acylated ghrelin increases glucose output from pig hepatocytes [30].

Interestingly, unacylated ghrelin, although devoid of any neuroendocrine action, is able to counteract the effects of acylated ghrelin on insulin secretion and glucose levels in humans [41]. This evidence led us to hypothesize that unacylated ghrelin would exert metabolic actions counterbalancing those of acylated ghrelin. To support this hypothesis, unacylated ghrelin inhibits glucose output and counteracts the stimulatory effect of both acylated ghrelin and glucagon on glucose output from pig hepatocytes [30]. Moreover, unacylated ghrelin abolishes the inhibitory effect of acylated ghrelin on the insulin response to glucose in hamster pancreatic β cells [personal unpubl. results].

Consistent with the hypothesis that acylated and unacylated ghrelin exert opposite influence on insulin secretion and glucose metabolism there is also evidence that transgenic rats specifically overexpressing unacylated ghrelin show a trend toward reduction of insulin and glucose levels [42]. In the same study, unacylated ghrelin has been shown to also have an anorectic action [42].

This picture would therefore suggest a diabetogenic action of acylated ghrelin counterbalanced by unacylated ghrelin [2, 30, 42]. There are, however, discrepancies that need to be explained. For instance, it has been reported that low total ghrelin levels are independently associated with type 2 diabetes mellitus, insulin secretion and sensitivity and blood pressure. Based on this association, the authors concluded that ghrelin might have a role in the etiology of diabetes mellitus type 2 and in the regulation of blood pressure homeostasis [43]. Notably, however, this report did dot allow to distinguish between acylated and unacylated circulating ghrelin levels and only this opportunity, now available [44], will allow to understand the real potential link between ghrelin and the metabolic syndrome. Then, it is also surprising that, differently from that observed in rats, low ghrelin secretion in type 1 diabetes mellitus is not modified by insulin replacement and normoglycemia [45]. This would be of particular interest considering personal unpublished results indicating that both acylated and unacylated ghrelin are able to markedly protect β cell viability from apoptosis induced by serum restriction as well as by the combination of interleukin and TNF.

Very recently, the complexity of this picture has been enhanced by the isolation of a new hormone from the stomach, derived from the same ghrelin gene and named obestatin [46]. Obestatin treatment in rats results in a reduction of food intake and body weight gain via activation of the orphan G protein-coupled receptor GPR39 [46].

These results would therefore describe the existence of a very complex system in which two products from the same gene are generated and, via a distinct receptor, seem to have an opposite metabolic effect. Moreover, when ghrelin is generated, a posttranscriptional acylation process may lead, in its turn, to two different bioactive ghrelin forms with different, sometimes opposite, metabolic effects.

Ghrelin, the Adipose Tissue and Lipid Metabolism

The adipogenetic action of ghrelin reflects its orexigenic action as well as its central modulatory effect on energy expenditure [2, 18]. However, there is increasing evidence indicating that ghrelin also directly acts on the adipose tissue [2, 47].

Ghrelin expression in the adipose tissue has never been demonstrated. On the other hand, the expression of GHS-R 1a mRNA in adipocytes has been found to be low by some authors and absent by others [2, 48, 49]. However, adipocytes also possess specific receptors for ghrelin independently from its acylation [2, 49] that are, by definition, GHS-R non-type 1a.

Consistent with the presence of binding sites specific for both acylated and unacylated ghrelin on adipocytes, it has been shown that both the forms of ghrelin directly promote adipogenesis in rat bone marrow adipocytes [49]. The direct adipogenetic effect of ghrelin has been confirmed in another study where acylated ghrelin was found able to inhibit the lipolytic action of a β-adrenergic agonist on rat adipocytes [50]. The direct effect of acylated ghrelin on proliferation, differentiation and apoptosis of adipocytes has also been shown by other authors who hypothesized that the ghrelin system would play a role in regulating fat cell number [51]. This action would be mediated by the activation of MAPK and PI3K/Akt pathways [51]. Inhibition of adipogenesis has also been reported in an adipocyte cell line overexpressing ghrelin [47].

Focusing on the modulatory effect of ghrelin on adipocyte function, we found that unacylated ghrelin is as active as its acylated form and more than a synthetic GHS such as hexarelin and MK-0677 in inhibiting isoproterenol-induced lipolysis in rat epididymal adipocytes in vitro [50]. That both acylated and unacylated ghrelin affect lipolysis further supports the hypothesis that adipocyte cell viability and function are likely to be mediated by a ghrelin receptor that is not the GHS-R1a [49]. It remains to be clarified why acylated and unacylated ghrelin modulate adipocyte function in the same way while they seem to have opposite influences on both pancreatic β cell secretion and glucose metabolism.

Ghrelin and Atherogenesis

Given the remarkable impact of the ghrelin system on glucose and lipid metabolism, its potential role in atherogenesis and cardiovascular functions has been obviously investigated.

Ghrelin and GHS-R 1a expression and binding sites have been demonstrated in cardiac tissues as well as in endothelial cells [1, 31, 52]. A GHS-R subtype that binds both acylated and unacylated ghrelin has been demonstrated in both these tissues [53].

Such a heterogeneous pattern of receptors for ghrelin and synthetic GHS within the cardiovascular system most likely explains the distinct cardiovascular actions exerted by these molecules [1, 2, 31, 52].

Acylated ghrelin as well as synthetic GHS improve cardiac performance in rats after myocardial infarction, protect against diastolic dysfunction of myocardial stunning in isolated perfused rabbit heart and enhance left-ventricular contractility in pigs with dilated cardiomyopathy [31, 53]. Moreover, the administration of peptidyl GHS increases left ventricular ejection fraction in normal volunteers as well as in hypopituitary patients with severe GHD and even in patients with ischemic dilated cardiomyopathy [31, 53]. These effects seem to be mediated by the GHS-R1a.

On the other hand, both acylated and unacylated ghrelin reduce the tension developed at low frequencies of guinea pig cardiac papillary muscles [54]. Moreover, either acylated or unacylated ghrelin as well as synthetic GHS protect cardiomyocytes and endothelial cells from serum withdrawal-, doxorubicin- and FAS ligand-induced apoptosis [31, 55]. This action is therefore mediated by a GHS-R non-type 1a.

Significant up-regulation of ghrelin and GHS receptors has been detected in atherosclerotic coronary arteries and in ventricular myocardium of patients with ischemic or dilated cardiomyopathy [31, 56]. On the other hand, synthetic peptidyl GHS exert protective action against cardiovascular damage in aged rats as well as in GHD rats with postischemic ventricular dysfunction [31].

This latter finding would indicate that the protective action of peptidyl GHS is not mediated by a ghrelin receptor and is consistent with the existence of a cardiac receptor specific for synthetic peptidyl GHS only [31, 53]. Indeed, cardiac tissues express a GHS-R type with a larger molecular mass (84 kDa) than that of GHS-R type 1a [31]; the predicted amino acid sequence of this putative GHS-R is identical to that of CD36, a multifunctional B-type scavenger receptor involved in foam cell formation and atherogenesis that is expressed in cardiomyocytes and microvascular endothelial cells [31, 52]. The activation of CD36 in perfused hearts by peptidyl GHS such as hexarelin (but not by MK-0677 or ghrelin) has been shown able to elicit an increase in coronary perfusion pressure in a dose-dependent manner and this effect is lacking in hearts from CD36-null mice and hearts from spontaneously hypertensive rats genetically deficient in CD36 [31, 52, 57]. Based on this evidence, it has been suggested that CD36 may mediate the coronary vasospasm seen in hypercholesterolemia and atherosclerosis.

Based on evidence that peptidyl GHS bind CD36, a receptor involved in scavenging of oxidized low-density lipoproteins (oxLDL) in monocytes/macrophages leading to foam cell formation, other studies focused on the hypothesis that GHS may interfere with CD36 function/expression in monocytes/macrophages, thereby hampering the uptake of oxLDL by macrophages and foam cells development. Under treatment with GHS significant reduction of the atherosclerotic lesion area (ranging from 28 to 47%) was observed in ApoE null mice. This was associated with a decrease in total plasma cholesterol and non-HDL cholesterol while HDL cholesterol tended to increase. GHS treatment also reduced ox-LDL-induced peritoneal macrophage accumulation by 39% and this reduction was paralleled with the increase of the expression of genes PPARγ, LXRa and ABCA1. In this study, it was therefore demonstrated that: (1) GHS protects ApoE null mice from developing fatty streaks lesions; (2) the anti-atherosclerotic effect of GHS is mediated through the negative modulation of CD36 expression in macrophages and the increase of expression of genes involved in the cellular cholesterol removal; (3) the protective effect of GHS is associated with favorable modulation in the plasma lipid profile [58]. Apart from clinical implications, these findings emphasize how relevant the direct metabolic actions exerted by ghrelin as well as by GHS receptors whose ligand is not always ghrelin are likely to be; in fact, a GHS receptor (i.e. CD36) but not ghrelin mediates a remarkable antiatherogenic action [52].

Conclusions

The ghrelin story was born with synthetic GHS following the perspective of orally active molecules able to replace rhGH for the treatment of GH deficiency and/or as anabolic treatment for anti-aging intervention. However, after its isolation, the demonstration of ghrelin's orexigenic action opened new clinical perspectives hypothesizing that ghrelin analogues act as agonists or antagonists in the treatment of cachexia, eating disorders and obesity. Evidence that the ghrelin KO mouse is not an anorectic dwarf dampened the enthusiasm in this field meantime describing an important difference between ghrelin and leptin (whose KO mouse has a clear obese phenotype). This, however, does not definitely rule out the potential importance of ghrelin. Ghrelin exerts pleiotropic actions consistent with the widespread distribution of ghrelin and GHS-R expression. Some ghrelin actions are independent of its acylation in serine 3 and this evidence strongly suggests the existence of GHS-R subtypes that would play a major role in peripheral metabolism. Although the orexigenic action of ghrelin in itself predicts impact on metabolism, ghrelin has also been shown to be able to directly act at the level of the endocrine pancreas, liver and adipose tissue thus modulating glucose and lipid metabolism. As previously stated, the complexity of this picture has been recently enhanced by the isolation of obestatin, a new hormone deriving from the same ghrelin gene, but apparently endowed with different metabolic actions mediated by the activation of a distinct receptor. The challenge is now the understanding the role of the ghrelin/obestatin system in the physiological control of metabolic functions and in the metabolic syndrome.

Acknowledgements

Personal studies reported in this review have been supported by MIUR (Rome, Italy), Theratechnologies (Canada) and the SMEM Foundation (Turin, Italy).

References

1 Van Der Lely AJ, Tschop M, Heiman ML, Ghigo E: Biological, physiological, pathophysiological, and pharmacological aspects of ghrelin. Endocr Rev 2004;25:426–457.
2 Ghigo E, Broglio F, Arvat E, Maccario M, Papotti M, Muccioli G: Ghrelin: more than a natural GH secretagogue and/or an orexigenic factor. Clin Endocrinol (Oxf) 2005;62:1–17.
3 Kojima M, Hosoda H, Date Y, Nakazato M, Matsuo H, Kangawa K: Ghrelin is a growth-hormone-releasing acylated peptide from stomach. Nature 1999;402:656–660.
4 Gottero C, Broglio F, Prodam F, Destefanis S, Bellone S, Benso A, Gauna C, Arvat E, Van Der Lely AJ, Ghigo E: Ghrelin: a link between eating disorders, obesity and reproduction. Nutr Neurosci 2004;7:255–270.
5 Janssen JA, Van Der Lely AJ, Lamberts SW: Is there a role of ghrelin in preventing catabolism? J Endocrinol Invest 2004;27:400–403.
6 Sun Y, Wang P, Zheng H, Smith RG: Ghrelin stimulation of growth hormone release and appetite is mediated through the growth hormone secretagogue receptor. Proc Natl Acad Sci USA 2004;101:4679–4684.
7 Sun Y, Ahmed S, Smith RG: Deletion of ghrelin impairs neither growth nor appetite. Mol Cell Biol 2003;23:7973–7981.
8 Sun Y, Asnicar M, Taffet G, Smith RG: Phenotype of Ghrelin Receptor Knockout Mice. The Endocrine Society 86th Annual Meeting, 2005, pp 3–47.
9 Ariyasu H, Takaya K, Tagami T, Ogawa Y, Hosoda K, Akamizu T, Suda M, Koh T, Natsui K, Toyooka S, Shirakami G, Usui T, Shimatsu A, Doi K, Hosoda H, Kojima M, Kangawa K, Nakao K: Stomach is a major source of circulating ghrelin, and feeding state determines plasma ghrelin-like immunoreactivity levels in humans. J Clin Endocrinol Metab 2001;86:4753–4758.
10 Cummings DE, Purnell JQ, Frayo RS, Schmidova K, Wisse BE, Weigle DS: A preprandial rise in plasma ghrelin levels suggests a role in meal initiation in humans. Diabetes 2001;50:1714–1719.
11 Paik KH, Jin DK, Song SY, Lee JE, Ko SH, Song SM, Kim JS, Oh YJ, Kim SW, Lee SH, Kim SH, Kwon EK, Choe YH: Correlation between fasting plasma ghrelin levels and age, body mass index (BMI), BMI percentiles, and 24-hour plasma ghrelin profiles in Prader-Willi syndrome. J Clin Endocrinol Metab 2004;89:3885–3889.
12 Tschop M, Weyer C, Tataranni PA, Devanarayan V, Ravussin E, Heiman ML: Circulating ghrelin levels are decreased in human obesity. Diabetes 2001;50:707–709.
13 Overduin J, Frayo RS, Grill HJ, Kaplan JM, Cummings DE: Role of the duodenum and macronutrient type in ghrelin regulation. Endocrinology 2005;146:845–850.
14 Barkan AL, Dimaraki EV, Jessup SK, Symons KV, Ermolenko M, Jaffe CA: Ghrelin secretion in humans is sexually dimorphic, suppressed by somatostatin, and not affected by the ambient growth hormone levels. J Clin Endocrinol Metab 2003;88:2180–2184.
15 Arosio M, Ronchi CL, Gebbia C, Cappiello V, Beck-Peccoz P, Peracchi M: Stimulatory effects of ghrelin on circulating somatostatin and pancreatic polypeptide levels. J Clin Endocrinol Metab 2003;88:701–704.

16 Tannenbaum GS, Epelbaum J, Bowers CY: Interrelationship between the novel peptide ghrelin and somatostatin/growth hormone-releasing hormone in the regulation of pulsatile growth hormone secretion. Endocrinology 2003;144:967–974.
17 Broglio F, Gottero C, Van KP, Prodam F, Destefanis S, Benso A, Gauna C, Hofland L, Arvat E, Van Der Lely AJ, Ghigo E: Acetylcholine regulates ghrelin secretion in humans. J Clin Endocrinol Metab 2004;89:2429–2433.
18 Konturek SJ, Konturek JW, Pawlik T, Brzozowski T: Brain-gut axis and its role in the control of food intake. J Physiol Pharmacol 2004;55:137–154.
19 Chan JL, Bullen J, Lee JH, Yiannakouris N, Mantzoros CS: Ghrelin levels are not regulated by recombinant leptin administration and/or three days of fasting in healthy subjects. J Clin Endocrinol Metab 2004;89:335–343.
20 Bruder ED, Jacobson L, Raff H: Plasma leptin and ghrelin in the neonatal rat: interaction of dexamethasone and hypoxia. J Endocrinol 2005;185:477–484.
21 Giordano R, Picu A, Pagotto U, De IR, Bonelli L, Prodam F, Broglio F, Marafetti L, Pasquali R, Maccario M, Ghigo E, Arvat E: The negative association between total ghrelin levels, body mass and insulin secretion is lost in hypercortisolemic patients with Cushing's disease. Eur J Endocrinol 2005;153:535–543.
22 Giordano R, Picu A, Broglio F, Bonelli L, Baldi M, Berardelli R, Ghigo E, Arvat E: Ghrelin, hypothalamus-pituitary-adrenal (HPA) axis and Cushing's syndrome. Pituitary 2004;7:243–248.
23 Otto B, Tschop M, Heldwein W, Pfeiffer AF, Diederich S: Endogenous and exogenous glucocorticoids decrease plasma ghrelin in humans. Eur J Endocrinol 2004;151:113–117.
24 Tsolakis AV, Portela-Gomes GM, Stridsberg M, Grimelius L, Sundin A, Eriksson BK, Oberg KE, Janson ET: Malignant gastric ghrelinoma with hyperghrelinemia. J Clin Endocrinol Metab 2004;89:3739–3744.
25 Chanoine JP, Wong AC: Ghrelin gene expression is markedly higher in fetal pancreas compared with fetal stomach: effect of maternal fasting. Endocrinology 2004;145:3813–3820.
26 Vartiainen J, Poykko SM, Raisanen T, Kesaniemi YA, Ukkola O: Sequencing analysis of the ghrelin receptor (growth hormone secretagogue receptor type 1a) gene. Eur J Endocrinol 2004;150: 457–463.
27 Smith RG, Leonard R, Bailey AR, Palyha O, Feighner S, Tan C, McKee KK, Pong SS, Griffin P, Howard A: Growth hormone secretagogue receptor family members and ligands. Endocrine 2001;14:9–14.
28 Kojima M, Kangawa K: Ghrelin: structure and function. Physiol Rev 2005;85:495–522.
29 Deghenghi R, Broglio F, Papotti M, Muccioli G, Ghigo E: Targeting the ghrelin receptor: orally active GHS and cortistatin analogs. Endocrine 2003;22:13–18.
30 Gauna C, Delhanty PJ, Hofland LJ, Janssen JA, Broglio F, Ross RJ, Ghigo E, Van Der Lely AJ: Ghrelin stimulates, whereas des-octanoyl ghrelin inhibits, glucose output by primary hepatocytes. J Clin Endocrinol Metab 2005;90:1055–1060.
31 Benso A, Broglio F, Marafetti L, Lucatello B, Seardo MA, Granata R, Martina V, Papotti M, Muccioli G, Ghigo E: Ghrelin and synthetic growth hormone secretagogues are cardioactive molecules with identities and differences. Semin Vasc Med 2004;4:107–114.
32 Kageyama H, Funahashi H, Hirayama M, Takenoya F, Kita T, Kato S, Sakurai J, Lee EY, Inoue S, Date Y, Nakazato M, Kangawa K, Shioda S: Morphological analysis of ghrelin and its receptor distribution in the rat pancreas. Regul Pept 2005;126:67–71.
33 Rindi G, Necchi V, Savio A, Torsello A, Zoli M, Locatelli V, Raimondo F, Cocchi D, Solcia E: Characterisation of gastric ghrelin cells in man and other mammals: studies in adult and fetal tissues. Histochem Cell Biol 2002;117:511–519.
34 Wierup N, Sundler F: Ultrastructure of islet ghrelin cells in the human fetus. Cell Tissue Res 2005;319:423–428.
35 Prado CL, Pugh-Bernard AE, Elghazi L, Sosa-Pineda B, Sussel L: Ghrelin cells replace insulin-producing beta cells in two mouse models of pancreas development. Proc Natl Acad Sci USA 2004;101:2924–2929.
36 Murata M, Okimura Y, Iida K, Matsumoto M, Sowa H, Kaji H, Kojima M, Kangawa K, Chihara K: Ghrelin modulates the downstream molecules of insulin signaling in hepatoma cells. J Biol Chem 2002;277:5667–5674.

37 Reimer MK, Pacini G, Ahren B: Dose-dependent inhibition by ghrelin of insulin secretion in the mouse. Endocrinology 2003;144:916–921.
38 Dezaki K, Hosoda H, Kakei M, Hashiguchi S, Watanabe M, Kangawa K, Yada T: Endogenous ghrelin in pancreatic islets restricts insulin release by attenuating Ca^{2+} signaling in beta-cells: implication in the glycemic control in rodents. Diabetes 2004;53:3142–3151.
39 Egido EM, Rodriguez-Gallardo J, Silvestre RA, Marco J: Inhibitory effect of ghrelin on insulin and pancreatic somatostatin secretion. Eur J Endocrinol 2002;146:241–244.
40 Broglio F, Gottero C, Benso A, Prodam F, Destefanis S, Gauna C, Maccario M, Deghenghi R, Van Der Lely AJ, Ghigo E: Effects of ghrelin on the insulin and glycemic responses to glucose, arginine, or free fatty acids load in humans. J Clin Endocrinol Metab 2003;88:4268–4272.
41 Broglio F, Gottero C, Prodam F, Gauna C, Muccioli G, Papotti M, Abribat T, Van Der Lely AJ, Ghigo E: Non-acylated ghrelin counteracts the metabolic but not the neuroendocrine response to acylated ghrelin in humans. J Clin Endocrinol Metab 2004;89:3062–3065.
42 Ariyasu H, Takaya K, Iwakura H, Hosoda H, Akamizu T, Arai Y, Kangawa K, Nakao K: Transgenic mice overexpressing des-acyl ghrelin show small phenotype. Endocrinology 2005;146: 355–364.
43 Poykko SM, Kellokoski E, Horkko S, Kauma H, Kesaniemi YA, Ukkola O: Low plasma ghrelin is associated with insulin resistance, hypertension, and the prevalence of type 2 diabetes. Diabetes 2003;52:2546–2553.
44 Akamizu T, Shinomiya T, Irako T, Fukunaga M, Nakai Y, Nakai Y, Kangawa K: Separate measurement of plasma levels of acylated and desacyl ghrelin in healthy subjects using a new direct ELISA assay. J Clin Endocrinol Metab 2005;90:6–9.
45 Soriano-Guillen L, Barrios V, Lechuga-Sancho A, Chowen JA, Argente J: Response of circulating ghrelin levels to insulin therapy in children with newly diagnosed type 1 diabetes mellitus. Pediatr Res 2004;55:830–835.
46 Zhang JV, Ren PG, vsian-Kretchmer O, Luo CW, Rauch R, Klein C, Hsueh AJ: Obestatin, a peptide encoded by the ghrelin gene, opposes ghrelin's effects on food intake. Science 2005;310: 996–999.
47 Zhang W, Zhao L, Lin TR, Chai B, Fan Y, Gantz I, Mulholland MW: Inhibition of adipogenesis by ghrelin. Mol Biol Cell 2004;15:2484–2491.
48 Choi K, Roh SG, Hong YH, Shrestha YB, Hishikawa D, Chen C, Kojima M, Kangawa K, Sasaki S: The role of ghrelin and growth hormone secretagogues receptor on rat adipogenesis. Endocrinology 2003;144:754–759.
49 Thompson NM, Gill DA, Davies R, Loveridge N, Houston PA, Robinson IC, Wells T: Ghrelin and des-octanoyl ghrelin promote adipogenesis directly in vivo by a mechanism independent of the type 1a growth hormone secretagogue receptor. Endocrinology 2004;145:234–242.
50 Muccioli G, Pons N, Ghe C, Catapano F, Granata R, Ghigo E: Ghrelin and des-acyl ghrelin both inhibit isoproterenol-induced lipolysis in rat adipocytes via a non-type 1a growth hormone secretagogue receptor. Eur J Pharmacol 2004;498:27–35.
51 Kim MS, Yoon CY, Jang PG, Park YJ, Shin CS, Park HS, Ryu JW, Pak YK, Park JY, Lee KU, Kim SY, Lee HK, Kim YB, Park KS: The mitogenic and antiapoptotic actions of ghrelin in 3T3-L1 adipocytes. Mol Endocrinol 2004;18:2291–2301.
52 Cao JM, Ong H, Chen C: Effects of ghrelin and synthetic GH secretagogues on the cardiovascular system. Trends Endocrinol Metab 2006;17:13–18.
53 Muccioli G, Broglio F, Tanabra E, Ghigo E: Known and unknown growth hormone secretagogue receptors and their ligands; in Ghigo E (ed): Ghrelin: More than Simply a Natural GH Secretagogue and/or an Orexigenic Factor. Boston, Kluwer Academic, 2004, pp 27–48.
54 Bedendi I, Alloatti G, Marcantoni A, Malan D, Catapano F, Ghe C, Deghenghi R, Ghigo E, Muccioli G: Cardiac effects of ghrelin and its endogenous derivatives des-octanoyl ghrelin and des-Gln14-ghrelin. Eur J Pharmacol 2003;476:87–95.
55 Baldanzi G, Filigheddu N, Cutrupi S, Catapano F, Bonissoni S, Fubini A, Malan D, Baj G, Granata R, Broglio F, Papotti M, Surico N, Bussolino F, Isgaard J, Deghenghi R, Sinigaglia F, Prat M, Muccioli G, Ghigo E, Graziani A: Ghrelin and des-acyl ghrelin inhibit cell death in cardiomyocytes and endothelial cells through ERK1/2 and PI 3-kinase/AKT. J Cell Biol 2002;159: 1029–1037.

56 Katugampola SD, Pallikaros Z, Davenport AP: [^{125}I-His(9)]-ghrelin, a novel radioligand for localizing GHS orphan receptors in human and rat tissue: up-regulation of receptors with athersclerosis. Br J Pharmacol 2001;134:143–149.
57 Bodart V, Febbraio M, Demers A, McNicoll N, Pohankova P, Perreault A, Sejlitz T, Escher E, Silverstein RL, Lamontagne D, Ong H: CD36 mediates the cardiovascular action of growth hormone-releasing peptides in the heart. Circ Res 2002;90:844–849.
58 Ong H, Tremblay-Lebeau A, Febbraio M, Silverstein RL, Avallone R, Iken K, Wang Y, Bujold K, Demers A, Sirois M, Deghenghi R, Marleau S: Growth hormone-releasing peptides as inhibitors of fatty streaks formation: a new therapy for atherosclerosis. Endocrine Society 85th Annual Meeting, 2003, pp 2–223.

E. Ghigo, MD
Division of Endocrinology and Metabolism
Department of Internal Medicine, University of Turin
C.so Dogliotti 14
IT–10126 Torino (Italy)
Tel. +39 011 6334317/6963156, Fax +39 011 6647421, E-Mail ezio.ghigo@unito.it

Arzt E, Bronstein M, Guitelman M (eds): Pituitary Today: Molecular, Physiological and Clinical Aspects. Front Horm Res. Basel, Karger, 2006, vol 35, pp 115–128

Regulating of Growth Hormone Sensitivity by Sex Steroids: Implications for Therapy

Ken K.Y. Ho, James Gibney, Gudmundur Johannsson, Troels Wolthers

Pituitary Research Unit, Garvan Institute of Medical Research and Department of Endocrinology, St. Vincent's Hospital, Sydney, Australia

Abstract

Growth hormone (GH) is an important regulator of body composition, reducing body fat by stimulating fat oxidation and enhancing lean body mass by stimulating protein accretion. The emergence of differences in body composition between the sexes during puberty suggests sex steroids modulate the action of GH. Work from our laboratory have investigated the influence of estrogens and androgens on the metabolic actions of GH in human adults. The liver is an important site of physiological interaction as it is a sex steroid responsive organ and a major target of GH action. Estrogen, when administered orally impairs the GH-regulated endocrine and metabolic function of the liver via a first-pass effect. It reduces circulating IGF-I, fat oxidation and protein synthesis, contributing to a loss of lean and a gain of fat mass. These effects occur in normal and in GH-deficient women and are avoided by transdermal administration of physiological doses of estrogen. In contrast, studies in hypopituitary men indicate that testosterone enhances the metabolic effects of GH. Testosterone alone stimulates fat oxidation and protein synthesis, both of which are enhanced by GH. Studies in GH deficiency adults have consistently reported women to be less sensitive to GH than men. In summary, estrogens and androgens exert divergent effects on the action of GH. The results provide an explanation for sexual dimorphism in body composition in adults and the gender-related response to GH replacement in hypopituitary subjects. In the management of hypopituitarism, estrogens should be administered by the parenteral route in women and testosterone be replaced in men to optimize the benefits of GH replacement.

The action of growth hormones (GH) is influenced by several factors. Growth in children is influenced by environmental and systemic factors, nutritional and endocrine status. Chronic systemic illness impairs growth as is observed in children with chronic pulmonary, cardiac, renal and hematological disorders. As growth is an anabolic process, adequate nutrition is essential for optimal growth. Malnutrition is the commonest cause of short stature worldwide. Untreated or poorly controlled type I diabetes is a functional form of malnutrition causing growth failure which is reversed by improved metabolic control. The mechanisms of insulin interaction with GH in growth regulation are complex and have been the subject of intensive study. Thyroid hormones are also important regulators of growth and exert positive effects on the secretion and the action of GH [1, 2]. There is strong evidence that sex hormones interact with the GH system as exemplified by the physical growth and somatic development that occurs in concert with enhanced levels of sex steroids and GH during puberty. Studies in adults have revealed that GH plays an important role in regulating substrate metabolism and body composition [3]. GH stimulates protein synthesis and lipolysis, metabolic actions which result in a gain of lean body mass and a loss of body fat. While growth is stimulated during puberty, differences in body composition between sexes also emerge at this time. However, GH secretion is increased to a comparable level in both sexes during the growth spurt. These observations suggest that the gender-emergent changes in body composition arise from the influence of female and male hormones on the GH/IGF-I system. We here review work undertaken in our laboratory on the modulation by estrogens and testosterone on GH action.

Estrogen and GH Action

There is strong evidence that estrogen regulates GH secretion in the human female. Serum estradiol and GH levels are significantly correlated in pubertal girls [4]. Estrogen priming in prepubertal girls augments the GH response to stimulation tests. GH secretion varies during the menstrual cycle with mean concentrations highest during the late follicular phase when estrogen concentration is highest [5]. Mean 24-hour GH and IGF-1 levels are lower in postmenopausal than in premenopausal women, suggesting that reduced activity of the somatotropic axis in the menopause may be secondary to estrogen deficiency [6]. Work investigating the effects of estrogen replacement on the GH-IGF-I axis has uncovered a strong dependency on how estrogen is administered.

Route of Estrogen Administration

This discovery was made from studies comparing the effects of oral and transdermal estrogen administration on circulating GH and on IGF-I [7].

Estrogen treatment in the menopause induces a significant route-dependent effect on the GH/IGF-1 axis. Administration of 20 μg ethinyl estradiol as an oral tablet and 100 mg 17β-estradiol delivered via a skin patch resulted in significant and comparable reductions in circulating levels of LH and FSH. Administration of oral ethinyl oestradiol resulted in a threefold increase in mean 24-hour GH concentrations. In contrast, transdermal administration of 17β-estradiol did not result in a significant change in mean 24-hour GH concentration [7]. Oral administration of ethinyl estradiol resulted in a uniform and significant reduction in mean IGF-1 levels. In contrast, transdermal estrogen delivery resulted in a small but significant increase in mean IGF-1 levels. Oral ethinyl estradiol but not transdermal estrogen increased circulating levels of GH binding protein, which is believed to be derived from proteolytic cleavage of the extracellular domain of the GH receptor [8].

In this study the types of estrogen were different. To address whether the reduction of IGF-1 and elevations of GH are specific properties of ethinyl estradiol or intrinsic to the oral route of administration, the effects of oral administration of ethinyl estradiol (20 μg), conjugated equine estrogen (Premarin 1.25 mg) and estradiol ester (estradiol valerate 2 mg) [9] were compared. The doses of the three formulations have previously been shown to have approximate systemic bioequivalence [10, 11].

All three estrogen formulations caused significant suppression of luteinitzing hormone (LH) and follicle-stimulating hormone (FSH), and elevations of the hepatic proteins SHBG and angiotensinogen. GHBP increased in parallel with these hepatic proteins. Each of the three estrogen formulations induced a fall in IGF-1 levels compared to baseline and corresponding elevations in mean 24-hour GH and GHBP concentrations (fig. 1). The increases in mean 24-hour GH during estrogen treatments were and inversely related to the fall in IGF-1 levels. Thus the uniform responses displayed by all three estrogen formulations administered by the oral route stand in contrast to those observed following transdermal delivery and strongly suggest that the reduction in IGF-1 levels is an intrinsic effect of oral estrogens [9]. The increase in GHBP concentration reflects another level of action of estrogen on the GH-IGF-1 axis. All three estrogen formulations increased GHBP in parallel with elevations in SHBG and angiotensinogen, both recognized as estrogen-sensitive hepatic proteins. As transdermal estrogen delivery did not influence GHBP levels, the data suggest that GHBP is an estrogen-sensitive hepatic protein and that the liver is a major source of GHBP in man [12]. IGFBP-3 and ALS, components of the IGF ternary complex are also reduced by oral and not transdermal estrogen [13]. ALS is derived exclusive from the liver which is also a major source of IGFBP-3.

We have proposed that the IGF-1 suppressive effect of oral estrogens arise as a consequence of exposure of the liver to high portal concentrations of

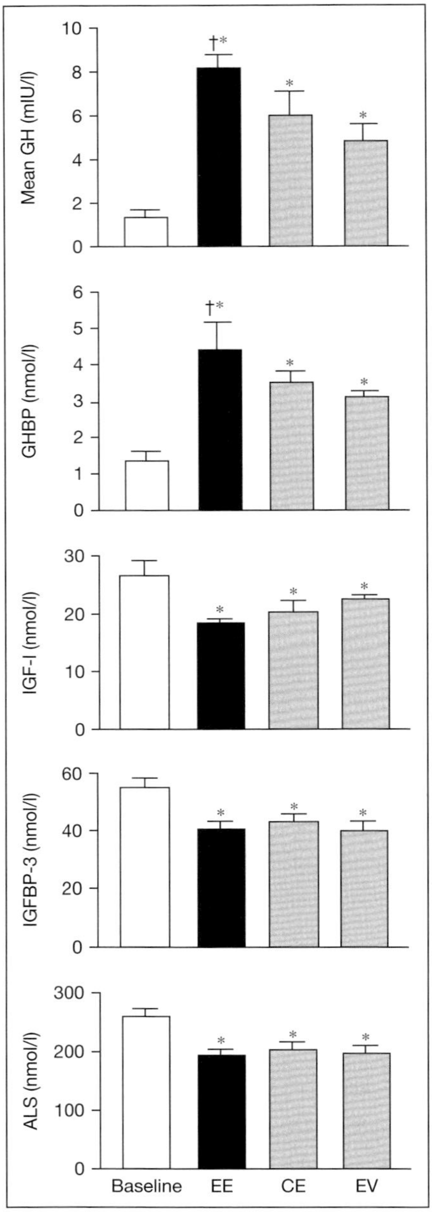

Fig. 1. Mean (± SE) IGF-1, mean 24-hour GH, GH-binding protein, IGFBP3 and ALS concentrations in 6 postmenopausal women before and during treatment with ethinyl estradiol (EE), conjugated equine estrogen (CE) and estradiol valerate (EV). *$p < 0.05$ vs. baseline; †$p < 0.05$ vs. EV. Adapted from Kelly et al. [9] and Kam et al. [13].

estrogen. This proposal is supported by evidence in the rat that estrogen inhibits hepatic mRNA expression [14]. It is known that this first-pass hepatic effect stimulates the synthesis of a number of hepatic proteins such as angiotensinogen, coagulation factors and lipoproteins, effects that are do not occur with the parenteral route. Because the fall in circulating IGF-1 occurred in the setting of enhanced GH secretion, the increase in GH concentration is a likely consequence of reduced negative feedback inhibition of IGF-1 on GH secretion [7].

Effects on GH Action

The dissociation induced by the oral route on the GH/IGF-I axis suggests that estrogen inhibited hepatic generation of IGF-I. We undertook a series of studies in GH-deficient and in postmenopausal women to investigate whether oral estrogen antagonized the metabolic and endocrine GH-regulated actions of the liver.

In GH-deficient women, we studied whether the biological effects of GH replacement was influenced by the route of estrogen administration. The effects of graded doses of GH (0.5, 1.0 and 1.5 U/day) on IGF-I, fat oxidation and whole body protein metabolism were compared in a group of women with hypopituitarism replaced with therapeutic doses of estrogen administered via the oral and transdermal routes [15]. At initiation of estrogen treatment, the mean IGF-1 level fell significantly with oral but not transdermal route of delivery. GH administration significantly increased IGF-I levels in a stepwise, dose-dependent manner during both estrogen treatments; however, mean IGF-I levels were significantly lower throughout the oral estrogen phase (fig. 2). Comparable IGF-I levels between estrogen treatments were obtained at a GH dosage approximately 1.0 IU higher during oral estrogen treatment. Postprandial lipid oxidation was significantly lower with oral estrogen treatment, both before and during GH administration, than with transdermal treatment. Protein synthesis was also lower during oral estrogen both before and during GH administration. Thus, estrogen at a therapeutic dose exerted significant route dependent effects on GH action in women with organic GH deficiency. When compared to the transdermal route, oral estrogen aggravates metabolic abnormalities of GH deficiency and attenuates the metabolic effects of GH therapy. Thus, in comparison to the transdermal route, oral estrogen markedly attenuated the metabolic effects of GH.

In post-menopausal women, we compared the effects of 6 months replacement with oral (Premarin 1.25 mg/day) and transdermal (Estraderm 100, 100 μg 17β-estradiol/day) on fat oxidation and on body composition [16] in a randomized cross-over study.

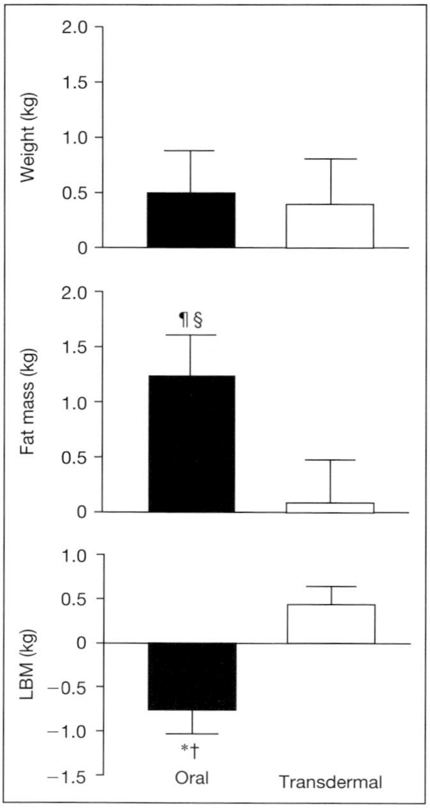

Fig. 2. Change (mean ± SE) in body weight (top panel), fat mass (middle panel) and lean body mass (LBM, lower panel) after 24 weeks of oral and transdermal estrogen treatment. ¶$p < 0.05$ oral vs. transdermal, *$p < 0.01$ oral vs. transdermal, †$p < 0.02$ vs. before oral estrogen treatment, §$p < 0.005$ vs. before oral estrogen treatment. Adapted from O'Sullivan et al. [37].

No differences in resting energy expenditure were observed between the routes of estrogen administration. When compared to transdermal estrogen, oral estrogen therapy resulted in an early but transient suppression of lipid oxidation and a reciprocal elevation of carbohydrate oxidation after a nutrient load. The finding of a suppressive effect of oral estrogen on fat oxidation extends the observations made in a young girl treated with high doses of ethinyl estradiol [17]. In this case report, we observed that oral ethinyl estradiol treatment with doses of 60, 100 and 200 μg/day produced a reversible, dose-dependent suppression of lipid oxidation associated with a reversible increase in fat. Lipid oxidation was reduced throughout the basal and postprandial states during ethinyl estradiol treatment in contrast to significant suppression occurring only during the postprandial phase in the present study. As the potency of ethinyl estradiol in the doses used is considerably higher than that of conjugated estrogen, the observations suggest that the inhibitory effect of oral estrogen on lipid oxidation is dose-dependent.

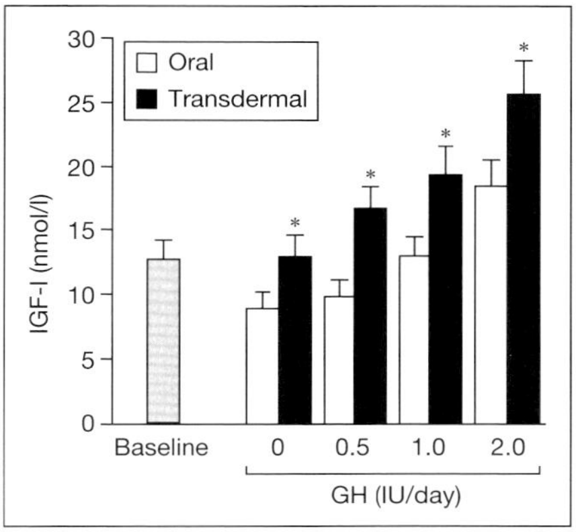

Fig. 3. Serum IGF-I concentrations (nmol/l, mean ± SE) before and during incremental dosages of GH (0.5, 1.0 and 2.0 IU/day) during oral and transdermal estrogen therapy. For conversion of GH dose to mg, divide by 3. *$p < 0.05$ by ANOVA, oral vs. transdermal.

No significant changes in body weight were observed between both routes of estrogen therapy after 6 months, nor did body weight change significantly with either treatment [16]. Mean bone mineral densities of the lumbar spine and femoral neck increased during oral and during transdermal estrogen therapy to a similar level. Significantly different effects on fat mass and lean body mass were observed between the two routes of estrogen therapy. When the effects of both routes of estrogen administration were compared, oral therapy lead to a significant increase in fat mass of 1.2 ± 0.5 kg, equivalent to a 5% change in body fat (fig. 3). This increase in fat mass arose from a significant increase occurring during 24 weeks of oral therapy with no significant change occurring during the transdermal estrogen phase. Oral estrogen therapy also induced a significant loss in lean body mass of 1.2 ± 0.4 kg ($p < 0.05$, equivalent to a 3% change) compared to that observed during transdermal estrogen therapy. This difference was accounted for by a significant decrease in lean body mass of 0.8 ± 0.3 kg with oral therapy and a small but nonsignificant increase in lean body mass with transdermal estrogen therapy (fig. 3). Thus, when compared to the transdermal route, oral estrogen therapy induced a significant decrease in lean body mass and a significant increase in whole body fat mass. The effects on fat oxidation and IGFI induced by the oral route of administration are opposite to the effects of GH and consistent with an antagonistic effect on GH action. These effects are even more significant considering that they have occurred in the setting of enhanced GH secretion [7, 9].

These studies in GH deficient and postmenopausal women provide compelling evidence that estrogen levels achieved in the portal circulation after ingestion of therapeutic doses of oral estrogen impair the GH-regulated function of the liver. In the laboratory, we have undertaken parallel in vitro studies to investigate the mechanism involved. We have recently reported that estrogen directly inhibits the function of the GH receptor by inducing the expression of proteins called SOCS which are negative regulators of GH signaling [18, 19].

Androgens and GH Action

There is strong evidence that androgens stimulate GH secretion. In males, GH secretion is reduced in hypogonadism and stimulated by androgen treatment, which also increases IGF-1 levels [20, 21]. Interestingly, the stimulatory effect of testosterone is dependent on aromatization to estrogen [22]. The question as to whether androgens amplify the action of GH has recently been addressed in our laboratory.

Studies in children suggest that both hormones interact positively to enhance growth and body composition [23, 24]. Testosterone enhances the growth of boys with hypogonadism, and with hypopituitarism during GH treatment [23, 24]. However, the effect of testosterone on somatic growth is poor in boys with hypopituitarism without concomitant GH treatment [23, 24]. In hypopituitary adults who are not receiving GH replacement, testosterone exerts no effect on circulating IGF-I [25]. These collective observations suggest that the growth promoting and anabolic effects of testosterone may be dependent on GH and possibly mediated in part by IGF-I.

Effects on Energy and Protein Metabolism

There is evidence that both hormones are necessary to exert an optimal anabolic effect. Even following adequate androgen replacement, LBM is reduced in GHD men [26]. The observation that the effects of GH replacement are more marked in men compared to women [27, 28] provides support that testosterone might enhance the anabolic effects of GH. The anabolic effects of GH are mediated by IGF-I, but whether IGF-I also plays a role in mediating the anabolic effects of testosterone is unknown. To investigate the interactions between these two anabolic hormones, we studied the independent and combined effects of GH and testosterone on IGF-I, energy expenditure and protein metabolism in hypopituitary men [29].

Treatment with GH alone significantly increased IGF-I into the normal range but did not alter plasma levels of testosterone, which remained in the hypogonadal range in all subjects. Addition of testosterone to GH increased testosterone into the normal range in all subjects, and resulted in a small but

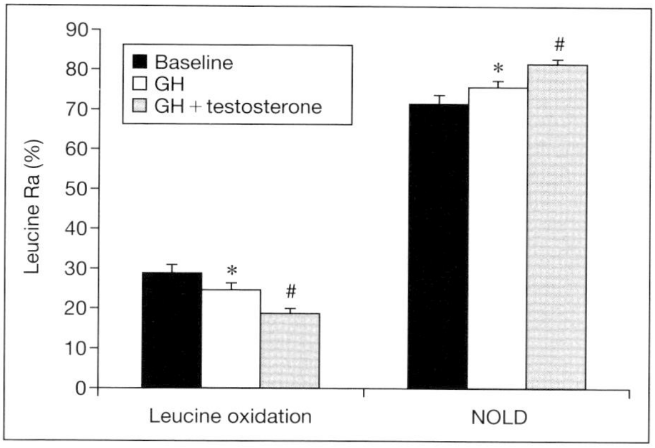

Fig. 4. Percentage rates of leucine oxidation and nonoxidative leucine disposal (NOLD), an index of protein synthesis in hypogonadal GH-deficient men at baseline, following treatment with GH alone and GH with testosterone *$p < 0.05$ vs. baseline. #$p < 0.05$ vs. baseline and vs. GH only. Adapted from Gibney et al. [29].

uniform increase in plasma IGF-I levels. Treatment with testosterone alone increased testosterone into the normal range but did not alter mean plasma levels of IGF-I. Compared with testosterone alone, treatment with combined GH and testosterone did not result in any further change in plasma testosterone, but increased IGF-I into the normal range. In summary, treatment with testosterone and GH normalized plasma levels of testosterone and IGF-I respectively, but testosterone increased IGF-I only during concomitant administration of GH [29].

GH alone stimulated mean resting energy expenditure by an average of 4.3%, and co-administration of testosterone enhanced resting energy expenditure by 8.6% over pretreatment. GH alone stimulated mean fat oxidation by 8.0% while addition of testosterone resulted in a further increase of over 30%. Using the whole-body leucine turnover technique to study effects on protein metabolism, we observed that GH treatment significantly reduced protein oxidation, and this effect was enhanced by combined treatment with testosterone. Similarly. GH-stimulated protein synthesis and the addition of testosterone enhanced this effect (fig. 4).

We also investigated the influence of testosterone on the antinatriuretic actions of GH by measuring changes in extracellular water volume in the hypopituitary men. GH significantly increased ECW volume, an effect which was enhanced by testosterone [30] (fig. 5). These changes occurred without a significant effect on components of the renin-angiotensin system nor on

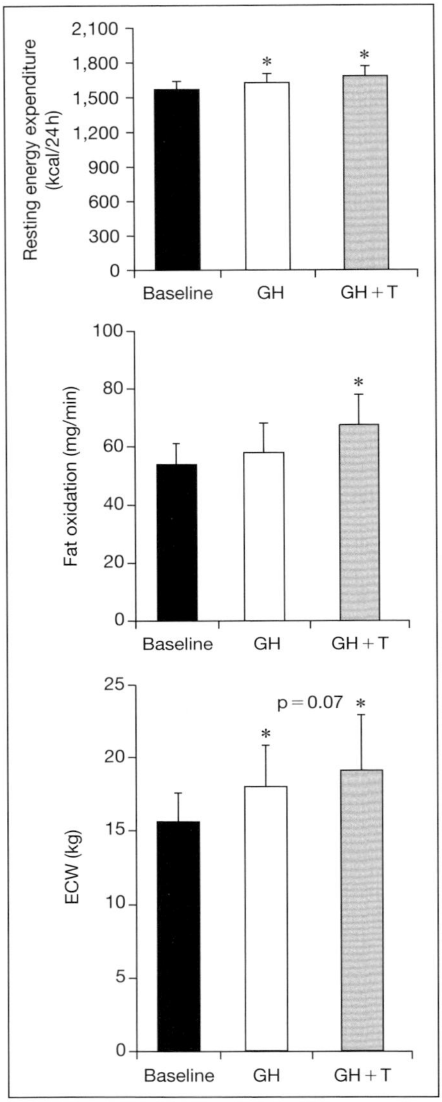

Fig. 5. Mean (± SE) resting energy expenditure (kcal/24 h), fat oxidation (mg/min) and extracellular water (ECW, kg) volume in 8 hypopituitary men at baseline, following treatment with GH alone, and GH with testosterone. *$p < 0.05$ vs. baseline. Adapted from Gibney et al. [29] and Johannsson et al. [30].

atrial-natriuretic peptide concentration. These observations suggest that the antinatriuretic effects of GH and its influence by androgens may be exerted at the renal tubular level. In summary, testosterone amplified the effects of GH on IGF-I, resting energy expenditure, fat oxidation, protein metabolism and extracellular water volume.

Implications

The divergent influence of estrogens and androgens on GH action provides an explanation for a consistent gender-related response observed in adults during GH replacement. IGF-I levels are lower in GH-deficient women than men replaced with similar doses of GH [31] and that the dose of GH required to maintain a comparable level of IGF-I is higher in women than men [32, 33]. Although these studies did not provide subgroup analysis on gonadal status and sex steroid treatments, it is likely that lower sensitivity to GH in women arose from concomitant oral estrogen treatment, the most widely used mode of replacement. Most of the hypogonadal men were androgen replaced. Using IGF-I as an endpoint for dose determination, Cook et al. [34] observed that GH requirements in women taking oral estrogens required at least twofold greater dosage of GH than those who did not. Janssen et al. [35] reported that switching from oral to transdermal estrogen therapy increased serum IGF-I by 30% during GH replacement therapy, an effect consistent with our own [15].

In a placebo-controlled 9-month trial, Burman et al. [27] observed that GH treatment induced a greater reduction in the proportion of total body fat and in abdominal fat mass in men than women. A large prospective study reported a progressive separation in fat mass, lean body and bone mass between men and women treated for up to 5 years with GH [33, 36]. In these studies, women actually received a larger weight-adjusted dose, indicating that they harbored an even greater degree of GH resistance than is apparent from the biochemical and body compositional change. Although the route of therapy was not specified, it can be assumed that the majority employed the oral route, as this was the usual mode of administration. As most of the hypogonadal men were also replaced with androgens in these studies, the positive regulatory effect of androgens is likely to be an additional mechanism explaining the clear sexual dimorphism in GH responsiveness in hypopituitary subjects.

Economic Considerations

Estrogens impair while androgens enhance the metabolic actions of GH. The effect of oral estrogen arises from a first-pass hepatic effect and is circumvented by parenteral administration. The route-dependent effect of estrogen has significant cost implications since a twofold higher dose of GH is required to achieve the same IGF-I level as that observed on transdermal therapy using standard estrogen replacement regimens. There are also important implications for men since GH acts optimally in the presence of testosterone. We recommend that estrogens be administered by a parenteral route in women and that testosterone be replaced in men with hypopituitarism to optimize the benefits of GH replacement.

Acknowledgement

Supported by Lilly, Pfizer and the National Health and Medical Research Council of Australia.

References

1 Chernausek SD, Underwood LE, van Wyk JJ: Influence of hypothyroidism on growth hormone binding by rat liver. Endocrinology 1982;111:1534–1538.
2 Chernausek SD, Underwood LE, Utiger RD, van Wyk JJ: Growth hormone secretion and plasma somatomedin-C in primary hypothyroidism. Clin Endocrinol 1983;19:337–344.
3 Ho KKY: Growth hormone deficiency in adults; in DeGroot LJ, Jameson JL (eds): De Groot's Textbook of Endocrinology. Philadelphia, Saunders, 2006, vol 1, pp 755–765.
4 Wennick JMB, Delemare-van-de Waal HA, Schoemaker R, Blaau G, van den Braken C, Schoemaker J: Growth hormone secretion patterns in relation to LH and estradiol secretion throughout normal female puberty. J Clin Endocrinol Metab 1990;124:129–135.
5 Faria ACS, Beckenstein LW, Booth RA, Vaccaro VA, Asplin CM, Veldhuis JD, Thorner MO: Pulsatile growth hormone release in normal women during the menstrual cycle. Clin Endocrinol 1992;36:591–596.
6 Ho KY, Evans WS, Blizzard RM, Veldhuis JD, Merriam GR, Samojlik E, Furlanetto R, Thorner MO: Effects of sex and age on the 24 hour secretory profile of GH secretion in man: importance of endogenous estradiol concentrations. J Clin Endocrinol Metab 1987;64:51–58.
7 Weissberger AJ, Ho KY, Lazarus L: Contrasting effects of oral and transdermal routes of estrogen replacement therapy on 24-hour growth hormone (GH) secretion, insulin-like growth factor 1 and GH binding protein in post-menopausal women. J Clin Endocrinol Metab 1991;72: 374–381.
8 Leung DW, Spencer SA, Cachaines G, Hammojnd RG, Collins C, Henzel WJ, Barnard R, Waters MJ, Wood WI: Growth hormone receptor and serum binding protein: purification, cloning and expression. Nature 1987;330:537–543.
9 Kelly JJ, Rajkovic IA, O'Sullivan AJ, Sernia C, Ho KKY: Effects of different oestrogen formulations on insulin-like growth factor-1, growth hormone and growth hormone binding protein in post-menopausal women. Clin Endocrinol 1993;39:561–567.
10 Maschak CA, Lobo R, Dozono TR, Eggena P, Nakamura RM, Brenner PF, Mishell DR: Comparison of pharmacokinetic properties of various estrogen formulations. Am J Obstet Gynecol 1982;144: 511–518.
11 Mandel FP, Geola FL, Lu J, Eggena P, Sambhi MP, Hershman JM, Judd HL: Biological effects of various doses of ethinyl estradiol. Am J Obstet Gynecol 1982;59:673–679.
12 Ho KY, Valiontis E, Waters MJ, Rajkovic IA: Regulation of growth hormone binding protein in man: comparison of gel chromatography and immunoprecipitation methods. J Clin Endocrinol Metab 1993;76:302–308.
13 Kam GYW, Leung KC, Baxter RC, Ho KKY: Estrogens exert route- and dose-dependent effects on insulin-like growth factor (IGF)-binding protein 3 and the acid-labile subunit of the IGF ternary complex. J Clin Endocrinol Metab 2000;85:1918–1922.

14 Murphy LJ, Freisen HG: Differential effects of estrogen and growth hormone on uterine and hepatic insulin-like growth factor-1 expression in the ovariectomised hypophysectomised rat. Endocrinology 1988;122:325–332.
15 Wolthers T, Hoffman DM, Nugent AG, Duncan MW, Umpleby M, Ho KKY: Oral estrogen therapy impairs the metabolic effects of growth hormone (GH) in GH deficient women. Am J Physiol 2001;281:E1191–E1196.
16 O'Sullivan AJ, Crampton LJ, Freund J, Ho KKY: The route of estrogen replacement therapy confers divergent effects on substrate oxidation and body composition in postmenopausal women. J Clin Invest 1998;102:1035–1040.
17 O'Sullivan AJ, Hoffman DM, Ho KKY: Estrogen, lipid oxidation and body fat. N Engl J Med 1995;333:669–670.
18 Leung KC, Doyler N, Ballesteros M, Sjogren K, Watts CKW, Low TH, Ross RJM, Ho KKY: Estrogen inhibits growth hormone (GH) signaling by suppressing GH-induced JAK2 phosphorylation: an effect mediated by SOC2. Proc Natl Acad Sci USA 2003;100:1016–1021.
19 Leung KC, Johannsson G, Leong GM, Ho KKY: Estrogen regulation of growth hormone action. Endocr Rev 2004;25:693–721.
20 Link K, Blizzard R, Evans W, Kaiser D, Parker M, Rogol A: The effect of androgens on the pulsatile release and the 24-hour mean concentration of growth hormone in peripubertal males. J Clin Endocrinol Metab 1986;62:159–164.
21 Liu L, Merriam GR, Sherins RJ: Chronic sex steroid exposure increases mean plasma growth hormone concentration and pulse amplitude in men with isolated hypogonadotropic hypogonadism. J Clin Endocrinol Metab 1987;64:58.
22 Weissberger AJ, Ho KKY: Activation of the somatotropic axis by testosterone in adult males: evidence for the role of aromatization. J Clin Endocrinol Metab 1993;76:1407–1412.
23 Zachmann M, Prader A: Anabolic and androgenic effect of testosterone sexually immature boys and its dependency on growth hormone. J Clin Endocr 1970;30:85–95.
24 Aynsley-Green A, Zachmann M, Prader A: Interrelation of the therapeutic effects of growth hormone and testosterone on growth in hypopituitarism. J Paediatr 1976;89:992–999.
25 Ip TP, Hoffman DM, Leung KC, Ho KKY: Do androgens regulate growth hormone binding protein in man? J Clin Endocrinol Metab 1995;80:1278–1282.
26 Hoffman DM, O'Sullivan A, Freund J, Ho KKY: Adults with growth hormone deficiency have abnormal body composition but normal metabolism. J Clin Endocrinol Metab 1995;80:72–77.
27 Burman P, Johansson AG, Siegbahn A, Vessby B, Karlsson FA: Growth hormone (GH)-deficient men are more responsive to GH replacement therapy than women. J Clin Endocrinol Metab 1997;82:550–555.
28 Svensson J, Stibrant Sunnerhagen K, Johannsson G: Five years of growth hormone replacement therapy in adults: age- and gender-related changes in isometric and isokinetic muscle strength. J Clin Endocrinol Metab 2003;88:2061–2069.
29 Gibney J, Wolthers T, Johannsson G, Umpleby AM, Ho KKY: Growth hormone and testosterone interact positively to enhance protein and energy metabolicm in hypopitutiary men. Am J Physiol 2005;289:E266–E271.
30 Johannsson G, Gibney J, Wolthers T, Leung K-C, Ho KKY: Independent and combined effects of testosterone and growth hormone on extracellular water in hypopituitary men. J Clin Endocrinol Metab 2005;90:3989–3994.
31 Johannsson G, Bjarnason R, Bramnert M, Carlsson LMS, Degerblad M, Manhem P, Rosén T, Thorén M, Bengtsson B-Å: The individual responsiveness to growth hormone (GH) treatment in GH-deficient adults is dependent on the level of GH binding protein, body mass index, age and gender. J Clin Endocrinol Metab 1996;81:1575–1581.
32 Drake WM, Parkinson C, Akker SA, Monson JP, Besser GM, Trainer PJ: Successful treatment of resistant acromegaly with a growth hormone receptor antagonist. Eur J Endocrinol 2001;145: 451–456.
33 Gotherstrom G, Svensson J, Koranyi J, Alpsten M, Bosaes I, Bengtsson B-Å, Johannsson G: A prospective study of 5 years of GH replacement therapy in GH-deficient adults: sustained effects on body composition, bone mass and metabolic indices. J Clin Endocrinol Metab 2001;86: 4657–4665.

34 Cook DM, Ludlam WH, Cook MB: Route of estrogen administration helps to determine growth hormone (GH) replacement dose in GH-deficient adults. J Clin Endocrinol Metab 1999;84: 3956–3960.
35 Janssen YJH, Helmerhorst F, Frolich M, Roelfsema F: A switch from oral (2 mg/day) to transdermal (50 g/day) 17β-estradiol therapy increases serum insulin-like growth factor levels in recombinant human growth hormone-substituted women with GH deficiency. J Clin Endocrinol Metab 2000;85:464–467.
36 Johannsson G, Grimby G, Sunnerhagen KS, Bengtsson BA: Two years of growth hormone treatment increases isometric and isokinetic muscle strength in GH-deficient adults. J Clin Endocrinol Metab 1997;82:2877–2884.
37 O'Sullivan AJ, Crampton L, Freund J, Ho KKY: Route of estrogen replacement confers divergent effects on energy metabolism and body composition in postmenopausal women. J Clin Invest 1998;102:1035–1040.

Ken K.Y. Ho
Garvan Institute of Medical Research
384 Victoria Road
Darlinghurst, Sydney NSW 2010 (Australia)
Tel. +61 2 9295 8482, Fax +61 2 9295 8481, E-Mail k.ho@garvan.org.au

Arzt E, Bronstein M, Guitelman M (eds): Pituitary Today: Molecular, Physiological and Clinical Aspects. Front Horm Res. Basel, Karger, 2006, vol 35, pp 129–134

Acromegaly: Molecular Expression of Somatostatin Receptor Subtypes and Treatment Outcome

Marcello D. Bronstein

Neuroendocrine Unit, Division of Endocrinology and Metabolism, Hospital das Clinicas, University of Sao Paulo Medical School, Sao Paulo, Brazil

Abstract

About a third of acromegalic patients are resistant to the currently commercially available somatostatin analogs (SA) octreotide and lanreotide. Such resistance is related to an overall reduction of somatostatin receptor (SSTR) density or to a differentiated expression of SSTR subtypes. There are five known SSTR subtypes. SSTR2 and SSTR5 are usually expressed in GH-secreting pituitary tumors, and both octreotide and lanreotide bind preferentially to SSTR2 and, to a lesser extent, to SSTR5. SA inhibitory effects on GH secretion and tumor cell proliferation can occur together or be dissociated events, depending on the tumor expression of SSTR subtypes involved in each mechanism. The development of specific somatostatin subtypes analogs, mainly for SSTR5, of a SSTR2–SSTR5 bispecific compound, and of a 'universal' analog with high affinity to SSTR1, 2, 3, and 5 showed preliminary, albeit promising results for the treatment of resistant somatotropic adenomas.

Somatostatin (SST) is a neuropeptide with important physiological regulations of neuroendocrine functions. Its action is mediated via 5 specific SST receptor subtypes (SSTR1–5), and exerts an inhibitory effect both in hormone secretion and tumor growth. The expression of SSTR subtypes among different types of pituitary adenomas and also among tumors secreting the same hormone is highly variable. SSTR2 and SSTR5 are the main subtypes involved in the regulation of GH secretion. About two thirds of acromegalic patients can be controlled by the commercially available somatostatin analogs (SA) octreotide and lanreotide, which bind with high affinity mainly to SSTR2 and with lesser extend to SSTR5 and SSTR3. The response to treatment is related to the expression of SSTR subtypes, so the best responders to octreotide (OCT) and LAN are acromegalic patients whose tumors express mainly SSTR2, whereas adenomas of acromegalics partially or completely resistant to SA mainly express other SSTR subtypes or are presented with very low or no expression of all of them. Therefore, the development of SSTR subtypes specific analogs could solve, at least partially, the problem of resistance to OCT and LAN in acromegaly.

SST is a cyclic peptide with important physiological regulations of neuroendocrine function across multiple biological systems [1]. It is produced by the hypothalamus, throughout the central nervous system, and in different peripheral organs, inhibiting hormone release and cell proliferation from the anterior pituitary, gastrointestinal tract and pancreas. This action is mediated via 5 specific SSTR subtypes, SSTR1, SSTR2, SSTR3, SSTR4 and SSTR5, belonging to the G protein-coupled receptor family, each of them codified by genes located in different chromosomes [1]. All five subtypes of SSTR are functionally coupled to inhibition of adenylyl cyclase via pertussis toxin-sensitive GTP-binding proteins. Some of the subtypes are also coupled to tyrosine phosphatase (SSTR1 and SSTR2) Ca^{2+} channels (SSTR2), Na^+/H^+ exchanger (SSTR1), PLA-2 (SSTR4), and MAP kinase (SSTR4) [1]. The expression of SSTR subtypes among different types of pituitary adenomas and also among tumors secreting the same hormone is highly variable. In normal and neoplastic human pituitary somatotropes cells SSTR1, SSTR2, SSTR3 and SSTR5 are prevalently expressed, SSTR2 and SSTR5 being the main subtypes involved in the regulation of GH secretion [2]. The short half-life of native SST, about 3 minutes, limits its use in clinical practice. The development of long-lasting SA as SC OCT, long-acting repeatable OCT (OCT-LAR), slow-release lanreotide (LAN-SR) and LAN autogel brought a great improvement to medical therapy of acromegaly [3–5]. However, as these SA bind with high affinity mainly to SSTR2 and with lesser extend to SSTR5 and SSTR3, the response to therapy depends on the individual tumor expression of SSTR subtypes [2]. SA have been shown to normalize GH and IGF-I in up to 60% of acromegalic patients, acting mainly through SSTR2 and SSTR5. However, with exceptions, tumor shrinkage is less frequent, and usually do not exceed 50% of size reduction [3]. The response to treatment is related to the expression of SSTR subtypes, so the best responders to OCT and LAN are acromegalic patients whose tumors express mainly SSTR2, whereas adenomas of acromegalics partially or completely resistant to SA mainly express other SSTR subtypes or are presented with very low or no expression of all of them. Therefore, the development of SSTR subtypes specific analogs could solve, at least partially, the problem of resistance to OCT and LAN in acromegaly. The premise for development of these analogs was based on studies demonstrating that a combination of ligands specific for SSTR2 and SSTR5 were synergistic for GH inhibition from GH-secreting adenoma.

Shimon et al. [6] studied in primary human GH-secreting pituitary adenoma cultures the effect on GH suppression of analogs preferential for SSTR2, including OCT and LAN, and novel compounds developed by Biomeasure Inc. with improved affinity for SSTR2 (BIM-23197), or SSTR5-selective compound (BIM-23268). They found that the novel analogs were 30–40% more potent than octreotide and lanreotide in suppressing GH. Heterologous analog combinations containing both SSTR2 and SSTR5-selective compounds in equimolar concentrations were more potent in decreasing GH than analogs used alone. This finding could be explained by evidences, based on fluorescence resonance energy transfer analysis, that activation by ligand induces SSTR dimerization, both homo- and heterodimerization with other members of the SSTR family, and that dimerization alters the functional properties of the receptor such as ligand binding affinity and agonist-induced receptor internalization and upregulation [7]. Additionally, Ben-Shlomo et al. [8] recently demonstrate in AtT-20 cells that SSTR5 modulates SSTR2 regulation of ACTH secretion.

Saveanu et al. [9], studying human GH adenoma cells in culture, showed that in adenomas of acromegalic patients highly sensitive to OCT, the mean SSTR2 expression was much higher than in tumors from partial OCT responders. Additionally, adenomas from OCT-responsive patients expressed SSTR2 and SSTR5 mRNA at an equivalent level. In contrast, adenomas from partial OCT responders with low SSTR2 mRNA expression expressed high levels of SSTR5 mRNA. In this series of experiments, the dose-response inhibition of GH release was examined with native SST14, the SSTR2-preferential compound, BIM-23197, and the SSTR5 preferential compound, BIM-23268. In cultures from the OCT-sensitive tumors the maximal GH suppression was obtained with SST14 and with BIM-23197, whereas the GH inhibition with BIM-23268 was much lower. In contrast, in the GH-secreting tumors partially responsive to octreotide, maximal GH suppression was equally achieved by SST14 and BIM-23268; BIM-23197 was slightly less potent than BIM-23268, indicating that in tumors partially responsive to OCT, the GH-suppressive effect is mediated through both the SSTR5 and SSTR2 subtypes. Additionally, in cell cultures from adenomas equally sensitive to the SSTR2- and SSTR5-preferential agonists (OCT partial responders), the dose-response inhibition of GH release by an SSTR2- and SSTR5-bispecific compound, BIM-23244 (Biomeasure Inc.) was similar to the inhibition induced by equimolar doses of BIM-23197 and BIM-23268.

A novel SST analog, named SOM230 (Novartis, Basel), with high binding affinity to SST receptor subtypes SSTR1, SSTR2, SSTR3 and SSTR5 was recently introduced [10]. His propriety to bind with high affinity to all SSTR with known functions, mainly to SSTR5, associated to a terminal elimination half-life of 27 h in humans, made this analog a potential candidate for treating

OCT and LAN-resistant acromegalic patients, as well as other pituitary and gastroenteropancreatic neuroendocrine tumors. Hofland et al. [11] showed that a higher number of GH-secreting adenomas cultures responded to SOM230, compared with OCT, suggesting that this novel analog has the potential to increase the number of patients controlled biochemically, both via SSTR2 and SSTR5. The same group compared in vivo the efficacy of a single dose of SOM 230 and OCT in 12 acromegalic patients. Both drugs showed comparable effects in eight patients, SOM230 was superior in three cases, and OCT was exhibited better result in one acromegalic [12]. More data is needed to confirm these initial results pointing to a better efficacy of SOM230 compared with OCT.

In acromegaly, the combination of STT and dopamine agonists has been shown to enhance suppression of GH secretion [13]. Both ligands bind to similar Gi protein receptors. Rocheville et al. [14] found that dopamine receptor D2R and somatostatin receptor SSTR5 interact physically through hetero-oligomerization to create a novel receptor with enhanced functional activity. Such direct intramembrane association defines a new level of molecular crosstalk between related G protein-coupled receptor subfamilies and provides a rational for the association of the two drugs. Saveanu et al. [15] demonstrate that a chimeric SST2-D2 molecule, BIM-23A387, had an enhanced potency in suppressing GH from cultures of responsive or partially responsive GH-secreting human pituitary adenomas compared to SSTR2 and D2DR specific analogs, either used individually or combined. Therefore, drugs based on chimeric SST2-D2 or SST2-SST5-D2 molecules could provide an additional tool for acromegaly treatment.

Tumor mass reduction during therapy with SA is an important event that reinforces medical treatment for acromegaly. Melmed et al. [16] compiling data from 14 studies encompassing 424 acromegalic patients showed that 36.6% (weighted mean percentage) of patients receiving primary SA therapy for acromegaly experienced a significant (>20%) reduction in tumor size. Data from our group [4] and from Colao et al. [17] showed 76 and 80% of significant tumor shrinkage, respectively. However, GH/IGF-1 control and tumor reduction by SA may occur separately. We observed an acromegalic male exhibiting an impressive shrinkage of 74% of his tumor volume during SA primary treatment without hormonal control. Interestingly enough, real-time RT-PCR assessment of his surgical removed GH-secreting adenoma showed only expression of the SSTR3 subtype which is linked to apoptosis, presumably by dephosphorylation-dependent conformational change in wild-type (wt) p53 as well as induction of Bax [18]. Danila et al. [19] observed a dissociation between in vitro effects of SST14 or LAN on tumor cell proliferation and the effects on GH secretion in human somatotrope tumors. These data suggest that inhibition of cell proliferation/induction of apoptosis may occur independently of effects on GH secretory pathways. Further studies are needed to clarify the mechanism of SST induced antiproliferative effects.

In summary, medical therapy with SA, primary or following other therapeutic modalities, becomes an important tool in the acromegaly treatment algorithm. The improvement in knowledge of SSTR subtypes role in the control of GH/IGF-1 secretion and tumor size is leading to the development of specific analogs which may overcome the resistance of one third of acromegalic patients to the commercially available OCT and LAN.

References

1 Patel YC, Greenwood MT, Panetta R, Demehyshyn L, Niznik H, Srikant CB: Mini review: the somatostatin receptor family. Life Sci 1995;57:1249–1265.

2 Shimon I: Somatostatin receptors in pituitary and development of somatostatin receptor subtype-selective analogs. Endocrine 2003;20:265–269.

3 Freda PU, Katznelson L, van der Lely AJ, Reyes CM, Zhao S, Rabinowitz D: Long-acting somatostatin analog therapy of acromegaly: a meta-analysis. J Clin Endocrinol Metab 2005;90:4465–4473.

4 Jallad RS, Musolino NR, Salgado LR, Bronstein MD: Treatment of acromegaly with octreotide-LAR: extensive experience in a Brazilian institution. Clin Endocrinol (Oxf) 2005;63:168–175.

5 Bronstein M, Musolino N, Jallad R, Cendros JM, Ramis J, Obach R, Leselbaum A, Catus F: Pharmacokinetic profile of lanreotide Autogel in patients with acromegaly after four deep subcutaneous injections of 60, 90 or 120 mg every 28 days. Clin Endocrinol (Oxf) 2005;63:514–519.

6 Shimon I, Yan X, Taylor JE, Weiss MH, Culler MD, Melmed S: Somatostatin receptor (SSTR) subtype-selective analogues differentially suppress in vitro growth hormone and prolactin in human pituitary adenomas: novel potential therapy for functional pituitary tumors. J Clin Invest 1997;100:2386–2392.

7 Rocheville M, Lange DC, Kumar U, Sasi R, Patel RC, Patel YC: Subtypes of the somatostatin receptor assemble as functional homo- and heterodimers. J Biol Chem 2000;275:7862–7869.

8 Ben-Shlomo A, Wawrowsky KA, Proekt I, Wolkenfeld NM, Ren SG, Taylor J, Culler MD, Melmed S: Somatostatin receptor type 5 modulates somatostatin receptor type 2 regulation of adrenocorticotropin secretion. J Biol Chem 2005;280:24011–24021.

9 Saveanu A, Gunz G, Dufour H, Caron P, Fina F, Ouafik L, Culler MD, Moreau JP, Enjalbert A, Jaquet P: Bim-23244, a somatostatin receptor subtype 2- and 5-selective analog with enhanced efficacy in suppressing growth hormone (GH) from octreotide-resistant human GH-secreting adenomas. J Clin Endocrinol Metab 2001;86:140–145.

10 Weckbecker G, Briner U, Lewis I, Bruns C: SOM230: a new somatostatin peptidomimetic with potent inhibitory effects on the growth hormone/insulin-like growth factor-I axis in rats, primates, and dogs. Endocrinology 2002;143:4123–4130.

11 Hofland LJ, van der Hoek J, van Koetsveld PM, de Herder WW, Waaijers M, Sprij-Mooij D, Bruns C, Weckbecker G, Feelders R, van der Lely AJ, Beckers A, Lamberts SW: The novel somatostatin analog SOM230 is a potent inhibitor of hormone release by growth hormone- and prolactin-secreting pituitary adenomas in vitro. J Clin Endocrinol Metab 2004;89:1577–1585.

12 van der Hoek J, de Herder WW, Feelders RA, van der Lely AJ, Uitterlinden P, Boerlin V, Bruns C, Poon KW, Lewis I, Weckbecker G, Krahnke T, Hofland LJ, Lamberts SW: A single-dose comparison of the acute effects between the new somatostatin analog SOM230 and octreotide in acromegalic patients. J Clin Endocrinol Metab 2004;89:638–645.

13 Cozzi R, Attanasio R, Lodrini S, Lasio G: Cabergoline addition to depot somatostatin analogues in resistant acromegalic patients: efficacy and lack of predictive value of prolactin status. Clin Endocrinol (Oxf) 2004;61:209–215.

14 Rocheville M, Lange DC, Kumar U, Patel SC, Patel RC, Patel YC: Receptors for dopamine and somatostatin: formation of hetero-oligomers with enhanced functional activity. Science 2000;288:154–157.
15 Saveanu A, Lavaque E, Gunz G, Barlier A, Kim S, Taylor JE, Culler MD, Enjalbert A, Jaquet P: Demonstration of enhanced potency of a chimeric somatostatin-dopamine molecule, BIM-23A387, in suppressing growth hormone and prolactin secretion from human pituitary somatotroph adenoma cells. J Clin Endocrinol Metab 2002;87:5545–5552.
16 Melmed S, Sternberg R, Cook D, Klibanski A, Chanson P, Bonert V, Vance ML, Rhew D, Kleinberg D, Barkan A: A critical analysis of pituitary tumor shrinkage during primary medical therapy in acromegaly. J Clin Endocrinol Metab 2005;90:4405–4410.
17 Colao A, Ferone D, Marzullo P, Cappabianca P, Cirillo S, Boerlin V, Lancranjan I, Lombardi G: Long-term effects of depot long-acting somatostatin analog octreotide on hormone levels and tumor mass in acromegaly. J Clin Endocrinol Metab 2001;86:2779–2786.
18 Sharma K, Patel YC, Srikant CB: Subtype-selective induction of wild-type p53 and apoptosis, but not cell cycle arrest, by human somatostatin receptor 3. Mol Endocrinol 1996;10:1688–1696.
19 Danila DC, Haidar JN, Zhang X, Katznelson L, Culler MD, Klibanski A: Somatostatin receptor-specific analogs: effects on cell proliferation and growth hormone secretion in human somatotroph tumors. J Clin Endocrinol Metab 2001;86:2976–2981.

Marcello D. Bronstein, MD
Av. 9 de Julho 3858
01406–100, Sao Paulo, SP (Brazil)
Tel. +55 11 3081 9120, Fax +55 11 3086 4363, E-Mail mdbronstein@uol.com.br

Arzt E, Bronstein M, Guitelman M (eds): Pituitary Today: Molecular, Physiological and Clinical Aspects. Front Horm Res. Basel, Karger, 2006, vol 35, pp 135–142

Gene Therapy in the Neuroendocrine System

C.B. Hereñú, G.R. Morel, M.J. Bellini, P.C. Reggiani, Y.E. Sosa, O.A. Brown, R.G. Goya

Institute for Biochemical Research-Histology B, Faculty of Medicine, National University of La Plata, La Plata, Argentina

Abstract

The implementation of experimental gene therapy in animal models of neuroendocrine diseases is an area of growing interest. In the hypothalamus, restorative gene therapy has been successfully implemented in Brattleboro rats, an arginine vasopressin (AVP) mutant which suffers from diabetes insipidus, and in Koletsky (fa^k/fa^k) and in Zucker (*fa/fa*) rats which have leptin receptor mutations that render them obese, hyperphagic and hyperinsulinemic. In the above models, viral vectors expressing AVP, leptin receptor b and proopiomelanocortin, respectively, were stereotaxically injected in the relevant hypothalamic regions. In rats, aging brings about a progressive degeneration and loss of hypothalamic tuberoinfundibular dopaminergic (TIDA) neurons, which are involved in the tonic inhibitory control of prolactin secretion and lactotropic cell proliferation. Stereotaxic injection of an adenoviral vector expressing insulin-like growth factor I corrected their chronic hyperprolactinemia and restored TIDA neuron numbers. Spontaneous intermediate lobe pituitary tumors in a retinoblastoma (Rb) gene mutant mouse were corrected by injection of an adenoviral vector expressing the human Rb cDNA and experimental prolactinomas in rats were partially reduced by intrapituitary injection of an adenoviral vector expressing the HSV1-thymidine kinase suicide gene. These results suggest that further implementation of gene therapy strategies in neuroendocrine models may be highly rewarding.

Gene therapy, the transfer of genetic material for therapeutic purposes, has undergone an explosive development in the last decade. Current molecular biology technology has made it possible to consider as feasible genetic manipulations that would have been deemed utopic not too long ago. Particularly important advances are being made in the improvement of gene transfer technology. Current efforts for the development of more efficient viral vectors focus on two main objectives, namely the achievement of cell-type specificity for transgene delivery and the design of vectors where, once the transgene is incorporated into the target cell, its expression can be regulated by small molecules.

Gene transfer to the central nervous system (CNS) poses significant challenges due to both the relative inaccessibility of the brain and spinal cord and the extraordinary complexity of CNS structures. On the other hand, this approach offers unique advantages for the effective delivery of therapeutic molecules to specific CNS regions affected by tumors, neurodegenerative processes or genetic defects.

Although a great deal of research efforts are been devoted to developing gene therapy strategies for neurological diseases [1], most of the work has been done in non-diencephalic brain regions. Much less work has been performed in the neuroendocrine system despite the unique advantages that it offers for the assessment of in vivo gene therapy strategies. In this article we will evidence these advantages by reviewing the core of documented studies in which in vivo gene therapy has been implemented in the neuroendocrine system of rodent models.

In vivo Gene Therapy in the Hypothalamus

Brattleboro Rat

The first animal model in which gene therapy was implemented in the hypothalamus was the Brattleboro rat. This mutant contains a single basepair mutation in the arginine vasopressin (AVP) gene which results in a highly conspicuous diabetes insipidus (DI) phenotype characterized by the production of large volumes of hypo-osmotic urine and compensatory polydipsia [2]. An adenoviral vector (AdAVP) encoding the rat AVP cDNA under the control of the cytomegalovirus (CMV) promoter was used in this animal model. When AdAVP was stereotaxically injected into the substantia innominata (a non-AVP producing hypothalamic area) of normal Wistar-Kyoto rats, expression of AVP mRNA was detected from 7 days to 6 months postinjection [3]. Injection of AdAVP into the supraoptic nucleus (SON, where part of the magnocellular neurons that produce AVP in normal animals lie) of Brattleboro rats resulted in substantial expression of AVP in magnocellular cells as well as in the presence of immunohistochemically detectable AVP in their axons projecting to the posterior pituitary. Measurement of urine output and urine osmolality showed that the symptoms of DI in the Brattleboro rats were significantly reduced for up to 4 months after injection of the viral vector [3]. An equine infectious anemia viral (EIAV) vector expressing AVP was injected into the hypothalamic supraoptic nuclei of Brattleboro rats resulting in expression of functional AVP peptide in magnocellular neurons. This was accompanied by a 100% recovery in water homeostasis as assessed by daily water intake, urine production, and urine osmolality lasting for at least 1 year [4].

Mutant Leptin Receptor Rat Models

Rat models of leptin receptor deficiency have been used to explore the restorative effect of gene therapy in the hypothalamus of these obese animals. In one study [5], leptin receptor defective Koletsky rats (fa^k/fa^k) were used. These animals received in the arcuate (ARC) nucleus a bilateral 0.5-μl injection containing either an adenoviral vector (Ad-*lepr*^{b}) expressing the signaling isoform of the leptin receptor $lepr^b$ (2.4×10^{12} pfu/ml) or an Ad vector (Ad-*lacZ*) expressing the β-galactosidase (β-gal) reporter gene (1.7×10^{12} pfu/ml). Sixteen days postinjection, Ad-*lepr*^{b} – but not Ad-*lacZ*-injected animals showed restored *lepr*^{b} mRNA levels (which are low in control mutants) in the ARC nucleus. Restored leptin receptor expression reduced both mean daily food consumption (13%) and body weight gain (33%). It also increased hypothalamic pro-opiomelanocortin (POMC) mRNA levels (65%) while decreasing neuropeptide Y mRNA levels by 30%, relative to Ad-*lacZ*-injected mutants. In contrast, Ad-*lepr*^{b} injection in the ARC nucleus of wild-type animals had no effect on the above parameters.

A similar experimental approach was implemented in Zucker (*fa/fa*) rats which are obese, hyperphagic and hyperinsulinemic as a consequence of a mutation in their leptin receptor. These animals received in the ARC nucleus a stereotaxic bilateral injection (3 μl/side; 1.28×10^9 particles) of either a recombinant adeno-associated viral (rAAV) vector expressing murine POMC (rAAV-POMC) or a rAAV expressing enhanced green fluorescent protein (rAAV-eGFP). At day 38 postinjection, the rAAV-POMC-injected animals showed a 4-fold increase in hypothalamic POMC expression and a 62% increase in melanocortin signaling (indicated by phosphorylation of the cAMP response element binding protein, CREB), relative to rAAV-eGFP-injected animals. A sustained reduction in food intake, weight gain and visceral adiposity was also observed in the experimental animals. POMC transgene delivery also increased the uncoupling of brown adipose tissue (BAT) protein. Circulating leptin, cholesterol and insulin were reduced in the rAAV-POMC-injected animals receptor [6].

Taken together these leptin receptor studies further demonstrate the suitability of the hypothalamus for the implementation of restorative gene therapy in mutant animal models.

Aging Female Rat

The dopaminergic (DA) neurons of the rat hypothalamus are grouped into two main areas, A_{12} and A_{14} [7, 8], with the DA perikarya of the A_{12} area being located in the ARC nucleus and in the periarcuate region [9]. The A_{14} DA neurons are mainly located within the paraventricular (PaV) and periventricular (PeV) nuclei, with a few scattered DA neurons in the anterior ventromedial (AVM) hypothalamic area [9, 10]. The A_{12} area and its corresponding axon terminals constitute the tuberoinfundibular dopaminergic (TIDA) system, whereas the A_{14} area and its fibers are known as the periventricular dopaminergic (PVDA) system. Both systems regulate prolactin (PRL) secretion by exerting a tonic inhibitory control on both PRL secretion and lactotrope proliferation [11]. In early studies, TIDA neuron function was reported to decline during aging in rats, with a marked reduction in hypothalamic, median eminence and neurointermediate lobe DA content in old (24–26 months) as compared with young (4 months) rats [12]. More significant, the rate of DA secretion into the hypophysial portal blood of aged (20–26 months) male and female rats was found to decline drastically when compared with young (2–4 months) counterparts [13, 14]. Although the above age-related alterations in hypothalamic DA secretion were ascribed to a functional decline of TIDA neurons rather than to TIDA neuron loss [12], more recent work in very old female rats (32 months) showed that at extreme ages, DA neuron loss occurs in the rat hypothalamus, particularly in the PaV nucleus [15]. The degeneration and loss of TIDA neurons during normal aging is associated, in the female rat, with progressive hyperprolactinemia [16] and the development of pituitary prolactinomas [17]. Interestingly, parkinsonian patients usually reveal functional alterations in the hypothalamo-PRL axis [18].

It should be pointed out that although a number of in vivo models have been developed for the study of the pathophysiology of Parkinson disease (PD) as well as for the assessment of new therapeutic strategies for this devastating pathology [19], they share a significant limitation namely, that the neurological lesions they study are caused by experimental manipulations rather than by aging, the only unequivocal risk factor for PD [20, 21]. In this context, the aging female rat emerges as an interesting model of spontaneous and progressive central DAergic dysfunction. Besides, the functional status of TIDA neurons can be readily and humanely monitored in the animals by measuring circulating PRL levels.

A protective effect of insulin-like growth factor-I (IGF-I) gene transfer has been reported in human DA cell cultures exposed to the toxin salsolinol [22]. In vivo, restorative IGF-I gene therapy was implemented in young (5 months) and senile (28 months) female rats, which received a single intrahypothalamic injection of 3×10^9 pfu of adenoviral vectors expressing either the reporter gene for β-gal or rat IGF-I (control and experimental group, respectively) and were sacrificed 17 days postinjection (fig. 1). In the young animals, neither vector modified serum PRL levels but in the IGF-I vector-injected senile rats a nearly full reversion of their hyperprolactinemic status was recorded. Morphometric analysis revealed a significant increase in the total number of tyrosine hydroxylase (TH)-positive cells in the hypothalamus of experimental as compared with control senile animals

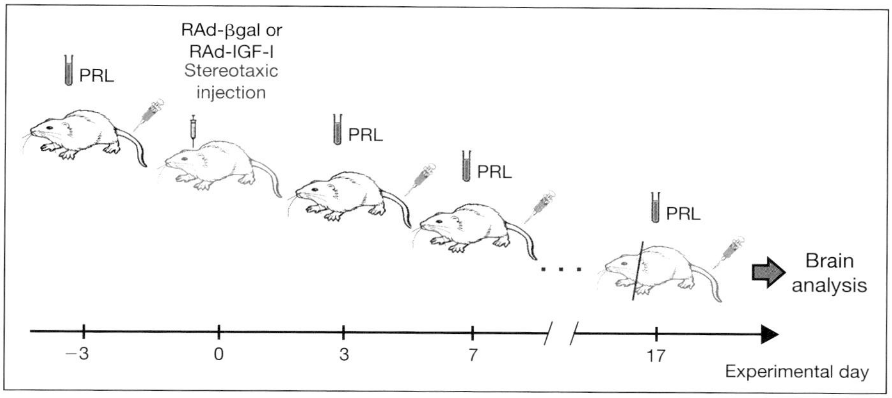

Fig. 1. Proposed experimental design for medium-term IGF-I gene therapy in the hypothalamus of senile female rats with TIDA neurodegeneration. All animals are bled from the tail veins at experimental days –3, 3, 7, 10, 14 and 17, in order to monitor serum PRL levels. RAd-β-gal (controls) or RAd-IGF-I (experimental) are stereotaxically injected in the hypothalamus of young and senile females on experimental day 0. At experimental day 17 (end of treatment) animals are sacrificed and brains removed for biochemical (e.g. transgenic protein content, dopamine and dopamine metabolites content, TH protein content), and morphometric assessment of hypothalami (e.g. TH+, NeuN+, CD11+ cell morphometric assessment).

[23, 24]. These results indicate that IGF-I gene therapy in senile female rats is highly effective for reversing their hypothalamic DAergic neurodegeneration.

In vivo Gene Therapy in the Pituitary Gland

Retinoblastoma Mutant Mouse

Another type of neuroendocrine model used to implement corrective gene therapy is that based on the transfer of a gene(s) with the ability to rescue the normal phenotype of pituitary tumor cells. This approach has been implemented in mice heterozygous for the retinoblastoma (Rb) tumor suppressor gene ($Rb^{+/-}$ mice). Such mice develop and succumb to characteristic pituitary intermediate lobe melanotrope tumors [25]. Transduction of tumor melanotropic cells with a recombinant adenoviral vector (rAd5.R.Rb) carrying the human Rb cDNA under the control of its own promoter showed a high level of efficiency both in vitro and in vivo [26]. Furthermore, intracranial delivery of this vector to mice carrying actively growing melanotropic tumors significantly reduced tumor growth and prolonged animal survival. Melanotropic tumor proliferative index and apoptotic rates were markedly lowered in the rAd5.R.Rb-treated animals, which also showed growth-inhibitory dopaminergic neuron reinnervation of melanotropic cells [26].

Suicide Gene Therapy in Experimental Pituitary Tumors

An adenoviral vector, RAdTK, harboring the herpes simplex type 1 thymidine kinase (HSV-1 TK) suicide gene under the control of the human CMV promoter has been used to transfer the TK gene to GH_3 and AtT_{20} rodent pituitary cell lines. Incubation of RAdTK-treated GH_3 and AtT_{20} cells with the prodrug ganciclovir (which after phosphorylation by viral TK becomes toxic) caused total destruction of the cultures [27]. In the same study, estrogen-induced rat prolactinomas were stereotaxically injected with the same RAdTK. Subsequent injection of the host animals with two daily i.p. doses of 25 mg ganciclovir/kg for 7 days partially succeeded in reducing AP tumor size and serum PRL levels.

Concluding Remarks

The implementation of experimental gene therapy in neuroendocrine models has revealed a number of advantages of this system for in vivo studies. The more relevant of these advantages is that the effectiveness and long-term duration of the treatment can be readily monitored by measuring hormone levels or other peripheral variables regulated by the neuroendocrine system. The age-related degeneration of TIDA neurons in female rats offers a unique animal model for the assessment of neuroprotective gene therapy strategies for PD. Further implementation of gene therapy strategies in neuroendocrine models may prove to be highly rewarding.

Acknowledgements

This work was supported in part by grants PICT13588 and PICT10663 from the (Argentine) National Agency for the Promotion of Science and Technology (ANPCyT) and grant NIH#1 R21 TW006665 from the Fogarty International Center and the National Institute on Aging (USA) to R.G.G. R.G.G. and O.A.B. are Research Career scientists of the Argentine Research Council (CONICET). C.B.H. is a doctoral fellowship recipient from CONICET, P.C.R. and G.R.M. are doctoral fellowship recipients from ANPCyT and M.J.B. is a postdoctoral fellowship recipient from NIH.

References

1 Lowenstein PR, Castro MG: Recent advances in the pharmacology of neurological gene therapy. Curr Opin Pharmacol 2004;4:91–97.

2 Valtin H: Genetic models of diabetes insipidus; in Windhager EE (ed): Handbook of Physiology. New York, Oxford University Press, 1992, pp 1281–1316.
3 Geddes BJ, Harding TC, Lightman SL, Uney JB: Long-term gene therapy in the CNS: reversal of hypothalamic diabetes insipidus in the Brattleboro rat by using an adenovirus expressing arginine vasopressin. Nat Med 1997;3:1402–1404.
4 Bienemann AS, Martin-Rendon E, Cosgrave AS, Glover CP, Wong LF, Kingsman SM, Mitrophanous KA, Mazarakis ND, Uney JB: Long-term replacement of a mutated nonfunctional CNS gene: reversal of hypothalamic diabetes insipidus using an EIAV-based lentiviral vector expressing arginine vasopressin. Mol Ther 2003;7:588–596.
5 Morton GJ, Niswender KD, Rhodes CJ, Myers MG Jr, Blevins JE, Baskin DG, Schwartz MW: Arcuate nucleus-specific leptin receptor gene therapy attenuates the obesity phenotype of Koletsky (fa^k/fa^k) rats. Endocrinology 2003;144:2016–2024.
6 Li G, Mobbs CV, Scarpace PJ: Central pro-opiomelanocortin gene delivery results in hypophagia, reduced visceral adiposity, and improved insulin sensitivity in genetically obese Zucker rats. Diabetes 2003;52:1951–1957.
7 Dahlström A, Fuxe K: Evidence for the existence of monoamine containing neurons in the central nervous system. I. Demonstration of monoamines in the cell bodies of brain stem neurons. Acta Physiol Scand 1964;62:1–55.
8 Tillet Y, Kitahama K: Distribution of central catecholaminergic neurons: a comparison between ungulates, humans and other species. Histol Histopathol 1998;13:1163–1177.
9 Kawano H, Daikoku S: Functional topography of the rat hypothalamic dopamine neuron systems: retrograde tracing and immunohistochemical study. J Comp Neurol 1997;265:242–253.
10 Goudreau JL, Lindley SE, Lookingland KJ, Moore KE: Evidence that hypothalamic periventricular dopamine neurons innervate the intermediate lobe of the rat pituitary. Neuroendocrinology 1992;56:100–105.
11 Ben-Jonathan N, Arbogast LA, Hyde JF: Neuroendocrine regulation of prolactin release. Prog Neurobiol 1989;33:399–477.
12 Porter JC, Nansel DD, Gudelsky GA, Reymond MJ, Pilotte NS, Foreman MM, Tilders FJH: Some aspects of hypothalamic and hypophysial secretion in aging rats. Peptides 1980;1:135–139.
13 Reymond MJ, Porter JC: Secretion of hypothalamic dopamine into pituitary stalk blood of aged female rats. Brain Res Bull 1981;7:69–73.
14 Gudelsky GA, Nansel DD, Porter JC: Dopaminergic control of prolactin secretion in the aging male rat. Brain Res 1981;204:446–450.
15 Sánchez HL, Silva LB, Portiansky EL, Goya RG, Zuccolilli GO: Impact of very old age on hypothalamic dopaminergic neurons in the female rat: a morphometric study. J Comp Neurol 2003;458:319–325.
16 Goya RG, Lu JKH, Meites J: Gonadal function in aging rats and its relation to pituitary and mammary pathology. Mech Age Dev 1990;56:77–88.
17 Cónsole G, Gómez-Dumm CL, Brown OA, Goya RG: Sexual dimorphism in the age-changes of the pituitary lactotrophs in rats. Mech Age Dev 1997;95:157–166.
18 Franceschi M, Camerlingo M, Perego L, Bottacchi E, Mamoli A: Tubero-infundibular dopaminergic function in Parkinson's disease. Eur Neurol 1988;28:117–119.
19 Shimohama S, Sawada H, Kitamura Y, Taniguchi T: Disease model: Parkinson's disease. Trends Mol Med 2003;9:360–365.
20 de Rijk MC, Breteler MM, Graveland GA, Ott A, Grobbee DE, van der Meche FG, Hofman A: Prevalence of Parkinson's disease in the elderly: the Rotterdam study. Neurology 1995;45: 2143–2146.
21 Mayeux R, Marder K, Cote LJ, Denaro J, Hemenegildo N, Mejia H, Tang MX, Lantigua R, Wilder D, Gurland B: The frequency of idiopathic Parkinson's disease by age, ethnic group and sex in northern Manhattan, 1988–1993. Am J Epidemiol 1995;142:820–827.
22 Shavali S, Ren J, Ebadi M: Insulin-like growth factor-1 protects human dopaminergic SH-SY5Y cells from salsolinol-induced toxicity. Neurosci Lett 2003;340:79–82.
23 Hereñú CB, Cristina C, Rimoldi OJ, Becú-Villalobos D, Goya RG: Construction and use of an adenoviral vector to express transgenic insulin-like growth factor I in the arcuate nucleus and substantia nigra of young and old rats (abstract). 12th International Congress of Endocrinology, Lisbon, 2004.

24 Hereñú CB, Cristina C, Rimoldi OJ, Becú D, Zucolilli G, Portiansky E, Goya RG: IGF-I gene therapy in the brain of senile rats with dopaminergic neurodegeneration (abstract). Invited Symposium, 18th World Congress of Gerontology, Rio de Janeiro, 2005.
25 Hu N, Gutsmann A, Herbert DC, Bradley A, Lee WH, Lee EY: Heterozygous Rb-1 delta 20/+mice are predisposed to tumors of the pituitary gland with a nearly complete penetrance. Oncogene 1994;9:1021–1029.
26 Riley DJ, Yu A, Lee W-H: Adenovirus-mediated retinoblastoma gene therapy suppresses spontaneous pituitary melanotroph tumors in $Rb^{+/-}$ mice. Nature Med 1996;2:1316–1321.
27 Windeatt S, Southgate TD, Dewey RA, Bolognani F, Perone MJ, Larregina AT, Maleniak TC, Morris ID, Goya RG, Klatzmann D, Lowenstein PR, Castro MG: Adenovirus-mediated herpes simplex virus type-1 thymidine kinase gene therapy suppresses oestrogen-induced pituitary prolactinomas. J Clin Endocrinol Metab 2000;85:1296–1305.

Rodolfo G. Goya
INIBIOLP, Faculty of Medicine, UNLP, CC 455
1900 La Plata (Argentina)
Tel. +54 221 425 6735, Fax +54 221 425 0924/425 8988
E-Mail goya@isis.unlp.edu.ar/rgoya@netverk.com.ar

Arzt E, Bronstein M, Guitelman M (eds): Pituitary Today: Molecular, Physiological and Clinical Aspects. Front Horm Res. Basel, Karger, 2006, vol 35, pp 143–156

Ectopic ACTH Syndrome

Andrea M. Isidori[a], *Gregory A. Kaltsas*[b], *Ashley B. Grossman*[c]

[a]DFM, Department of Pathophysiology, University 'La Sapienza' of Rome, Rome, Italy; [b]Department of Pathophysiology, National University of Athens, Athens, Greece; [c]Department of Endocrinology, St Bartholomew's Hospital, London, UK

Abstract

Ectopic adrenocorticotropin secretion (EAS) accounts for 10–15% of cases of Cushing's syndrome and comprises a spectrum of tumours from undetectable isolated lesions to widespread metastatic and aggressive malignancies. EAS is often associated with severe hypercortisolaemia causing hypokalaemia, diabetes, generalized infections, hypertension and psychotic reactions. Surgical resection of the primary lesion, achievable with a curative intent in about 40% of patients with EAS, is associated with complete remission in up to 80% of such cases. It is therefore mandatory to localize the source of ectopic ACTH hypersecretion in order to stage the disease and adopt optimal treatment modalities. Modern cross-sectional imaging techniques can identify the majority of the ACTH secreting lesions, either initially or at follow-up reassessment. However, in approximately 10–20% of patients with EAS, the source of ACTH hypersecretion remains occult in spite of extensive investigation and prolonged follow-up. In such cases, control of the hypercortisolemia can be achieved with long-term adrenolytic medication. When conditions require a prompt and definitive resolution of the hypercortisolaemia (i.e. as in pregnancy), bilateral adrenalectomy remains an alternative option. This review focuses on the clinical features, diagnostic pitfalls, management and long-term follow-up of the EAS based on the extensive experience of major referral centres.

Endogenous Cushing syndrome (CS) is a clinical state resulting from excessive cortisol secretion, characterized by the loss of the normal feedback mechanism of the hypothalamo-pituitary adrenal axis (HPA) and of the normal circadian rhythm of cortisol secretion [1]. The diagnosis of suspected Cushing's syndrome involves several stages: the confirmation of hypercortisolism, differentiation between adrenocorticotropin (ACTH)-independent and ACTH-dependent causes of Cushing's syndrome, and the differentiation between pituitary and ectopic sources of the ACTH (EAS) [2]. Since correction of cortisol hypersecretion achieved either by the selective removal of a pituitary microadenoma or of a solitary bronchial carcinoid tumour is associated with a high probability of cure, it is crucial to distinguish EAS from the much more common CD and, in the former, to make every effort to localize the source of ectopic ACTH [3, 4].

The EAS is associated with either highly malignant tumours, where circulating ACTH and cortisol levels are extremely elevated, the duration of symptoms is short and the clinical phenotype atypical, or with a variety of mainly neuroendocrine tumours exhibiting typical symptoms/signs of CS and biochemically resembling CD. The latter has also been referred as 'occult' EAS when the source of ectopic ACTH secretion is not obvious [5]. However, this term may no longer be suitable as recent improvements in endocrine dynamic testing, cross-sectional imaging and increasing attention to making an earlier diagnosis of CS have progressively changed the clinical spectrum of patients investigated for EAS. Therefore, we have recently proposed [6] a reclassification of the EAS as: *overt* when the tumoral source is detected following initial endocrine and radiological investigation; *covert* when it occurs in patients presenting with hypercortisolaemia in whom no ectopic source was detected at initial investigation, but became evident on subsequent evaluation or during prolonged follow-up, while *occult* EAS is reserved for patients with features of CS in whom all tests point towards an ectopic source, but the primary lesion cannot be identified even after prolonged and repeated follow-up.

Very recently, two large series of patients with EAS have been published: the first comes from the NIH [7] and the second from our own group from St. Bartholomew's Hospital, London [6]. These studies, sharing many similarities in the investigational protocols, results and conclusions, represent the most complete published collection of patients with the EAS to date. The findings of these studies are compared and contrasted here in order to highlight the current trends in the management of this rare but potentially treatable condition.

Source of ACTH Secretion

In the past, small cell lung cancers (SCLC) represented the predominant cause of EAS [8]. However, in recent surveys the preponderance of these tumours has been substantially reduced in favour of other histological types [5, 9]. This is likely to represent a change in referral patterns, rather than a true change in the incidence of these tumours, as the association between EAS and SCLC is now well recognized, and it is often diagnosed and managed by non-endocrinologists. Nevertheless, one recent analysis has revealed that the lung is still the most likely organ to harbour an ectopic source of ACTH hypersecretion from either SCLC (27%) or carcinoid tumours (21%) [10]. In our single institution series, the lung still remains the major organ (47.5%) harbouring an ACTH-secreting tumour, with the majority of cases being bronchial carcinoid tumours (30%) followed by SCLC (17.5%) [6]. Compared to other series, the prevalence of SCLC in our series was lower, probably because only patients in whom the features of CS represented the major complaint and reason for admission were included. In the NIH series, which actually included very few patients with SCLC [7], the lung was still the major source of ectopic ACTH hypersecretion in nearly 50% of tumours localized. Most importantly, in both series [6, 7], more than half of these tumours were found either in the thorax or neck, including thymic carcinoids and medullary thyroid cancers (MTC).

Bronchial carcinoid tumours exhibit considerable malignant potential and may share common histological features with SCLC, thus representing extremes along a tumour spectrum [11]. Although SCLC still remains one of the most common causes of EAS, only a minority of patients with SCLC, approximately 1.6–4.5%, demonstrate overt hypercortisolaemia, and this is probably associated with increased co-morbidity [12–14]. In our series we found that such patients suffer from complications related to the hypercortisolaemia such as pancreatitis, peritonitis and septicemia, and therefore control of the excess cortisol burden represents an important part of their management. Other sources of EAS were mostly related to tumors with neuroendocrine differentiation such as other non-lung carcinoids (22.5%), MTC (7.5%) and phaeochromocytoma (2.5%). Only 5% of cases with histological confirmation were derived from non-endocrine, non-pulmonary tumours. We also found that in spite of repeated evaluation and prolonged follow-up (up to 10 years) there are still a significant number of patients (12.5%) in whom the source of EAS cannot be found [6]. In the NIH series, this figure accounts for 19% of patients with EAS, but in that study the period of follow-up was shorter [7]. It should be emphasised that the diagnosis of *occult* EAS, according to the recently proposed definition, remains a diagnosis made by exclusion after prolonged investigation and long-term control of the hypercortisolaemia. In such cases, every effort should be made to localize the source of ectopic ACTH secretion, even if this requires several years, as cases of previously unsuspected sources of ACTH secretion have been identified during autopsy.

Clinical Presentation of the EAS

Describing the clinical symptoms/signs attributed to EAS from retrospective studies is occasionally difficult and may be misleading, as in many reports particular symptoms/signs may not be specifically mentioned and the absence of expected symptoms/signs is not always addressed. Our single institution series offers the advantage that the collection of relevant clinical information was performed on the basis of a consistent protocol in which specific symptoms/signs were assigned as either present or absent, thus eliminating individual variations of the interviewing physicians in data accrual [15]. The recorded clinical features of patients with EAS in our series were quite broad, resembling those found in patients with CD even in some patients with SCLC [5, 10]. It is likely that patients with SCLC, when evaluated at an earlier stage, may present in a similar manner to patients with other less malignant causes of EAS [10, 16]. However, patients with SCLC more commonly present with skin pigmentation and an absence of facial fullness or weight gain compared to other causes of EAS [6]. These differences are most likely due to the rapidity of onset and the severity of the hypercortisolaemia. On the contrary, no major differences were found between patients with occult and overt EAS suggesting that tumour mass (and consequently its detectability) does not necessarily predict the secretory characteristics of the tumour or the clinical features of the syndrome [6]. A recent single institutional study has shown a relatively high incidence (35%) of systemic infections in patients with EAS [17]. In our series, although cortisol levels were similar to those encountered in the previous studies, severe and/or life-threatening infections developed in approximately 15% of patients and this was related to the degree of hypercortisolaemia. Hypokalaemia was present in 70% of the patients in our series and this was again related to the degree of hypercortisolaemia [10, 16, 18]. In both the NIH and our series, a high prevalence of approximately 50% of psychiatric disorders was described [6, 7]; we found that these symptoms were a distinctive feature of the neuroendocrine tumours hypersecreting ACTH compared to SCLC, where the presence of pigmentation was more prominent. A further recent finding is that up to 50% of patients with EAS may present with osteoporosis or fractures [7].

Diagnosis of Ectopic ACTH Secretion

Establishing the diagnosis of CS in a timely and accurate fashion it is imperative as systemic hypercortisolaemia is associated with increased morbidity and mortality [19]. Once the diagnosis of CS has been confirmed, the differential diagnosis of ACTH-dependent causes is the second major challenge encountered by the physician. Although patients with the EAS tend to have higher ACTH levels compared to patients with CD, there is no clear distinction between these two entities. In a recent review, 64% of patients with the EAS were found to have ACTH levels greater than 200 pg/ml [10]. In our series, mean ACTH levels were 358.03 ± 90.28 pg/ml but half of the patients had ACTH below 200 pg/ml; as previously noted, patients with SCLC and EAS have higher ACTH levels compared to those with other causes of CS [16]. However, no statistically significant difference was found between patients with occult and overt EAS, once again indicating that small tumours may secrete large quantities of ACTH. Similarly, in the NIH series no difference was found in ACTH levels among patients with initially overt, diagnosed after 6 months (covert), or occult ACTH-secreting tumours. In addition, EAS may secrete biologically inactive fragments or precursors not detected in conventional ACTH assays, such as pro-opiomelanocortin (POMC) [20]. Therefore, it has been suggested that a high POMC/ACTH ratio may be regarded as a marker of tumour aggressiveness [21]. However, as CS is currently considered at an earlier stage, when the clinical and biochemical manifestations of the hypercortisolaemia are relatively mild, such alterations of the biochemical and endocrine markers may not be as commonly encountered, particularly in the diagnostically problematic occult bronchial carcinoid tumours [22].

Several tests have been proposed to establish the differential diagnosis of CS, with bilateral inferior petrosal sinus sampling (BIPSS) considered as the gold standard achieving a sensitivity and a specificity of over 95%. Traditionally, variations of the high-dose dexamethasone suppression test (HDDST) and/or the corticotropin-releasing hormone (CRH) test (either alone or in combination with desmopressin) have been used to discriminate patients with EAS from CD; tumours causing EAS are usually characterized by their unresponsiveness to glucocorticoid feedback, CRH or desmopressin [2, 19]. In our series, only 3.1% and 5.5% of patients with EAS showed a 'significant' response to the HDDST (using serum cortisol percentage change from base-line) and CRH test, respectively. A recent review, including a large number of patients with EAS, showed that 22–40% of patients demonstrate either serum or urinary 17-OHCS suppression to the HDDST [10, 16]; however, using serum rather than urinary cortisol or 17-OHCS as an indicator of adequate suppression a much higher sensitivity was obtained and we believe that measurement of serum cortisol represents the measurement of choice [1, 2]. However, patients with bronchial carcinoids show a much higher incidence of cortisol suppression, approximately 60% following the HDDST [2, 10]; variations of the formal HDDST were shown to exhibit a similar performance [23]. Additionally, approximately 10–15% of patients with EAS, using a variety of cut-off cortisol and/or ACTH criteria, have been shown respond to CRH administration [1, 2, 24]. Using desmopressin instead of CRH as a secretagogue a 40% false-positive rate in patients with EAS was obtained [25]. In our series desmopressin stimulation, either alone or in combination with CRH, did not improve the sensitivity or the specificity of the test. Indeed, recent studies have shown that the CRH test demonstrates a higher diagnostic accuracy than the HDDST [26, 27], while the accuracy of the HDDST is close to the pretest probability of having CD and therefore has a low predictive value [2, 22, 28]. Nevertheless, while each test on its own may be of relatively limited diagnostic accuracy, the combination of these two tests may aid in the diagnosis of EAS as a lack of response to both tests had a sensitivity of 100% and a diagnostic accuracy of 98% [2, 6, 29]. The combination of the various dynamic non-invasive endocrine tests can therefore be extremely helpful in difficult cases and should be considered when the clinical, biochemical and/or radiological findings are not consistent with the results obtained from the BIPSS [2]. The data from the NIH series show substantially similar findings with the percentage of EAS patients responding to dynamic testing ranging from 9 to 14% [7].

BIPSS is a powerful tool in excluding a pituitary source of ACTH hypersecretion, particularly if a microadenoma is not shown on MRI, but cannot identify the source of ectopic ACTH secretion [1]. Recent data have suggested that even when strict criteria, i.e. inferior petrosal sinus to peripheral (IPS:P) ACTH ratios of greater than 3 following CRH administration are applied [30], patients with BIPSS results suggestive of CD may still prove to have EAS [31, 32]. Secretion of CRH or co-secretion of CRH with ACTH, periodic secretion, and/or the presence of an ectopically situated pituitary adenoma (i.e. in the sphenoid sinus) might account at least in some cases for such results [5, 10, 31]. One patient in our series, who was later shown to have EAS, obtained a definite positive IPS ratio following CRH administration and underwent unnecessary transsphenoidal surgery (TSS). However, while such false-positive results suggestive of CD do undoubtedly occur they are extremely rare, and a positive BIPSS is a very strong indicator in favour of CD. On the other hand, several patients in whom the IPS:P ratio was suggestive of EAS were found to have CD after successful TSS [31]. It has been suggested that false-negative results after BIPSS are more common than previously appreciated, and that a negative BIPSS does not rule out a pituitary source of ACTH secretion as at least 30 patients with documented false-negative results have been described [3, 31, 33–36]. In our previous analysis we found a significant incidence of false-negative BIPSS (3%) [2]. Therefore, in the setting of a potential 'false-negative' BIPSS, it is very difficult to determine whether the result is truly false negative (i.e. the source of ACTH secretion being of pituitary origin), or whether the true ectopic source of ACTH secretion has not yet been found, until a further therapeutic/diagnostic procedure such as TSS is undertaken. This is particularly important as up to 10% of healthy individuals may harbour clinically non-significant pituitary micro-adenomas on MR scanning [37].

Two paradigmatic examples can be drawn from our series to illustrate the problems: the first patient showed a positive response to BIPSS while both the HDDST and CRH correctly suggested an ectopic source of ACTH. The patient underwent pituitary surgery and no abnormality was found; later investigation revealed the ectopic source of ACTH secretion to be a mesothelioma. The second patient showed a positive response to the CRH test and adequate suppression during the HDDST, but no gradient was found at BIPPS. All imaging studies were negative, so the patient underwent pituitary surgery that showed Crooke's changes, suggestive of hypercortisolism, but no pituitary adenoma. More than a year later the patient was found to have a bronchial carcinoid and was cured following a thoracotomy. Thus, while BIPSS is extremely helpful, and a positive response is very highly suggestive (but not pathognomonic) of CD, a negative response suggests that a careful search for an ectopic source should be undertaken, and that dynamic tests should be performed to obtain further useful information.

In both series of patients with EAS [6, 7], a variety of tumour markers were measured; however, they were proven to be of specific diagnostic value only in MTCs (calcitonin) and phaeochromocytomas (urinary catcholamines). Surprisingly, calcitonin and gastrin were the most commonly elevated tumour markers, regardless of tumour type. When tumour markers indicative of endocrine islet cell tumours were positive, such tumours were relatively large and radiologically evident. On the contrary, 5-HIAA levels were normal in the majority of patients, as neuroendocrine tumours arising from the embryonic foregut are usually deficient in the enzyme aromatic L-amino acid decarboxylase, therefore producing less serotonin but a greater tendency to peptide hormone production responsible for the EAS [38].

Localization of the ACTH-Secreting Tumour

After biochemical confirmation of EAS, optimal treatment includes localization and removal of the ACTH-secreting tumour [19, 22]. Localization of these tumours can occasionally be difficult and may require extensive and prolonged follow-up [1, 4]. The majority of SCLC are detected using plain chest radiology, CT and/or MR imaging [5, 10, 14]. In our series, all patients with SCLC except one had detectable lesions on either plain radiology or CT scanning of the chest, or both. One patient had an SCLC that was not found on initial CT imaging of the thorax, but which became evident when liver metastases were found. Bronchial carcinoids can be of relatively small size and thus be missed on conventional imaging; however, early application of 2–3 mm high-resolution CT scanning of the thorax, particularly with the new generation of multidetector CT, identifies the great majority of such cases [39]. In our series, CT and/or MRI failed to localize the ACTH-secreting tumours in approximately 12.5% of cases with EAS; this number is less than that reported in previous studies and may be related to the prolonged follow-up and the subsequent application of newer imaging modalities [10, 40]. It has been suggested that in such cases imaging with ^{111}In-octreotide can identify true occult ACTH-secreting tumours in the majority of patients with neuroendocrine tumours, including bronchial carcinoid, tumours [10, 41]. Even bronchial lesions less than 1 cm in diameter have been identified following imaging with ^{111}In-octreotide [10]. However, there has been some concern regarding the superiority of this method over CT scanning, as in a direct comparison ^{111}In-octreotide identified fewer lesions than CT [42], whereas no tumour was identified with ^{111}In-octreotide that was not demonstrated by CT [10, 40, 42]. In the NIH series, scintigraphy with ^{111}In-octreotide did not add further to the diagnostic information derived from other imaging modalities. However, there is currently no consensus as to whether scintigraphy with ^{111}In-octreotide, after failing to establish the diagnosis at first presentation, should be included in the follow-up of such patents, particularly after some more recent studies have shown that it may be of diagnostic utility [43, 44]. Pancreatic islet cell tumours associated with the EAS are usually large and have already metastasized to the liver by the time CS is diagnosed; no covert EAS secondary to an islet cell tumour was encountered according to one review [10]. Such tumours are usually functional and co-secrete other biologically active substances associated with characteristic clinical syndromes, and only occasionally is CS the presenting complaint [10].

Whole body catheterization studies (selective venous sampling from suspected sources of ectopic ACTH secretion) was helpful only in a minority of patients; in addition, this technique is technically difficult and overall seems to be of limited value even when imaging with modern diagnostic modalities is unrevealing [10, 45, 46]. Although POMC fragment analysis may be another tool to identify the ectopic sources of ACTH secretion, this awaits further clinical verification [47].

In summary, as suggested by the NIH group, we agree that a single imaging study may represent a falsely positive result, and therefore it is strongly recommended that CT, MRI, octreotide scanning (and maybe in the future PET scans) are used together to localize the source of ectopic ACTH.

Management of the EAS

Following the localization of the lesion responsible for ectopic ACTH secretion, the optimal management is surgical excision. Based on the recently published series this can be achieved with a curative intent in up to 40% of patients with EAS, and is associated with complete remission in more than 80% of such cases. The NIH series reports an overall 29% curative resection, as opposed to a 12% rate obtained at the Mayo Clinic [48]. All patients with EAS should receive treatment with either one or more adrenolytic medications to achieve eucortisolaemia as soon as the preliminary dynamic tests are completed. This is particularly important for patients with SCLC in whom CS is associated with increased co-morbidity, and therefore treatment with inhibitors of glucocorticoid production is initiated followed by cytotoxic chemotherapy [5]. In the NIH series, infections were present in 51% of patients with EAS. In general eucortisolism can be obtained in many patients without major side-effects; occasionally metyrapone and ketoconazole daily doses of up to 6,000 and 1,200 mg may be required. In cases where the long-term control of the hypercortisolaemia with drugs is unsatisfactory or contraindicated and in women with occult EAS who wish to become pregnant, bilateral adrenalectomy can be used to resolve the hypercortisolemia [5]. Bilateral adrenalectomy was required in 37 and 30% of cases in the NIH and our series, respectively. Alternatively, in severely ill adults [49, 50] and children [51], when metyrapone and ketoconazole are ineffective, etomidate infusion, combined with hydrocortisone therapy, can stabilize cortisol levels and is safe for the short-term control of severe hypercortisolaemia pre-operatively.

Survival analysis revealed that the prognosis of patients with EAS is positively correlated with the tumor histology. In our series, patients with SCLC had the worse prognosis, usually dying within 12 months following the diagnosis. Patients with other tumours with endocrine differentiation had a better prognosis, particularly those with bronchial carcinoid tumours compared to patients with other carcinoid tumours, MTC and phaeochromocytoma. In the subgroup of tumours with endocrine differentiation the presence of metastatic deposits was another important predictor of overall survival. The survival of patients with an occult source of ACTH, but adequate control of hypercortisolaemia, was very good.

Conclusions

The syndrome of ectopic ACTH secretion is a not uncommon cause of hypercortisolaemia accounting for approximately 15% of cases of endogenous CS. Two recent independent studies have presented the US and UK experience with this condition from tertiary referral centres, and showed very similar results [6, 7]. The first message that can be drawn is that nearly any neuroendocrine or non-endocrine tumour, including atypical tumours such as mesotheliomas, may be associated with EAS. However, bronchial carcinoid tumours are the commonest cause of EAS, mainly presenting with classical symptoms/signs of CS. In addition, as the diagnosis of hypercortisolaemia is now considered at an earlier stage, patients with SCLC may occasionally present with similar clinical findings.

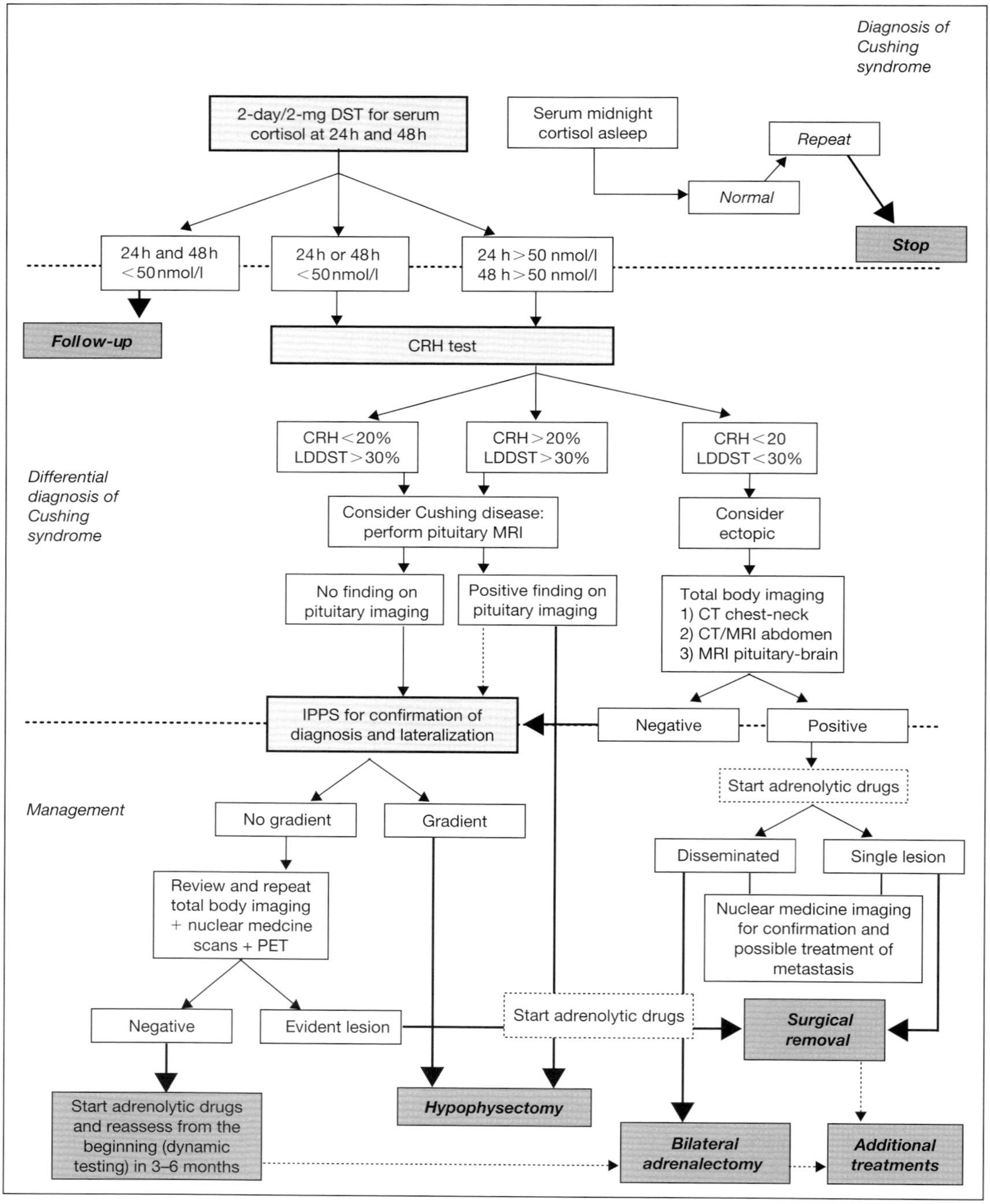
Diagnosis of Cushing syndrome
2-day/2-mg DST for serum cortisol at 24h and 48h
Serum midnight cortisol asleep
Repeat
Normal
Stop
24h and 48h <50nmol/l
24h or 48h <50nmol/l
24 h>50 nmol/l 48 h>50 nmol/l
Follow-up
CRH test
CRH<20% LDDST>30%
CRH>20% LDDST>30%
CRH<20 LDDST<30%
Differential diagnosis of Cushing syndrome
Consider Cushing disease: perform pituitary MRI
Consider ectopic
No finding on pituitary imaging
Positive finding on pituitary imaging
Total body imaging 1) CT chest-neck 2) CT/MRI abdomen 3) MRI pituitary-brain
IPPS for confirmation of diagnosis and lateralization
Negative
Positive
Start adrenolytic drugs
Management
No gradient
Gradient
Disseminated
Single lesion
Review and repeat total body imaging + nuclear medcine scans + PET
Nuclear medicine imaging for confirmation and possible treatment of metastasis
Negative
Evident lesion
Start adrenolytic drugs
Surgical removal
Start adrenolytic drugs and reassess from the beginning (dynamic testing) in 3–6 months
Hypophysectomy
Bilateral adrenalectomy
Additional treatments

The second message is that the size of the tumour does not correlate with the severity of symptoms, neither do the levels of ACTH with the aggressiveness of the lesion. Regarding diagnostic tests, no single endocrine test, including dynamic and invasive ones, is accurate enough to distinguish the ectopic from the pituitary sources of ACTH hypersecretion in every single case. BIPSS, when performed by a skilled team in referral centres, remains the gold standard procedure to confirm pituitary ACTH hypersecretion although its diagnostic accuracy is rather less than previously considered. High-resolution cross-sectional imaging is progressively gaining a pivotal role in the identification of the source of ectopic ACTH. Recent studies confirm that when the endocrinologist collaborates with the expert radiologist and the cross-sectional imaging is combined with the nuclear medicine scans, the majority of lesions can be identified either at initial investigation or during follow-up. A curative surgical attempt should be possible in more than 40% of patients with EAS, with a complete remission anticipated in more than 80% of these. Alternatively, a wait-and-see approach may avoid procedures associated with significant morbidity. In the past, many patients with clear discrepancies between biochemical tests and no radiologically evident tumour were considered to have CD and underwent TSS. We believe that this is no longer an evidence based approach: it is necessary to establish a diagnostic protocol, based on recent published data, modern imaging and a combination of dynamic tests [2, 31] and apply it to patients with EAS who fail to demonstrate a lesion after meticulous investigations. Our data indicate that a prolonged follow-up may be necessary before a covert tumour responsible for EAS becomes detectable, especially following the evolving improvement in imagining modalities. We suggest the development of a flow-chart for the management of the covert EAS, emphasizing that this is clearly open to future discussion and amendment (fig. 1). Following the diagnosis of EAS, hypercortisolemia needs to be adequately controlled, either medically or surgically, preferably pre-operatively to reduce the morbidity and mortality of the procedure. Bone fractures, infections and thrombosis are serious frequent complications in patients with EAS. Whenever surgery is not curative, a multidisciplinary approach should be considered including chemotherapy, radiotherapy and/or radionuclide treatment. Finally, bilateral adrenalectomy remains an alternative to long-term adrenolytic medication in patients with disseminated tumors or in patients with special needs or contraindications to these drugs.

Fig. 1. Suggested flow-chart for the diagnosis and management of Cushing syndrome.

References

1 Newell-Price J, Trainer P, Besser M, Grossman A: The diagnosis and differential diagnosis of Cushing's syndrome and pseudo-Cushing's states. Endocr Rev 1998;19:647–672.

2 Isidori AM, Kaltsas GA, Mohammed S, Morris DG, Jenkins P, Chew SL, Monson JP, Besser GM, Grossman AB: Discriminatory value of the low-dose dexamethasone suppression test in establishing the diagnosis and differential diagnosis of Cushing's syndrome. J Clin Endocrinol Metab 2003;88:5299–5306.

3 Kaltsas GA, Giannulis MG, Newell-Price JD, Dacie JE, Thakkar C, Afshar F, Monson JP, Grossman AB, Besser GM, Trainer PJ: A critical analysis of the value of simultaneous inferior petrosal sinus sampling in Cushing's disease and the occult ectopic adrenocorticotropin syndrome. J Clin Endocrinol Metab 1999;84:487–492.

4 Raff H, Findling JW: A physiologic approach to diagnosis of the Cushing syndrome. Ann Intern Med 2003;138:980–991.

5 Wajchenberg BL, Mendonca BB, Liberman B, Pereira MA, Carneiro PC, Wakamatsu A, Kirschner MA: Ectopic adrenocorticotropic hormone syndrome. Endocr Rev 1994;15:752–787.

6 Isidori AM, Kaltsas GA, Pozza C, Frajese V, Newell-Price J, Reznek RH, Jenkins P, Monson JP, Grossman AB, Besser GM: The ectopic adrenocorticotropin syndrome: clinical features, diagnosis, management and long-term follow-up. J Clin Endocrinol Metab 2006;91:371–377.

7 Ilias I, Torpy DJ, Pacak K, Mullen N, Wesley RA, Nieman LK: Cushing's syndrome due to ectopic corticotropin secretion: twenty years' experience at the National Institutes of Health. J Clin Endocrinol Metab 2005;90:4955–4962.

8 Imura H, Matsukura S, Yamamoto H, Hirata Y, Nakai Y: Studies on ectopic ACTH-producing tumors. II. Clinical and biochemical features of 30 cases. Cancer 1975;35:1430–1437.

9 Limper AH, Carpenter PC, Scheithauer B, Staats BA: The Cushing syndrome induced by bronchial carcinoid tumors. Ann Intern Med 1992;117:209–214.

10 Beuschlein F, Hammer GD: Ectopic pro-opiomelanocortin syndrome. Endocrinol Metab Clin North Am 2002;31:191–234.

11 Warren WH, Faber LP, Gould VE: Neuroendocrine neoplasms of the lung: a clinicopathologic update. J Thorac Cardiovasc Surg 1989;98:321–332.

12 Shepherd FA, Laskey J, Evans WK, Goss PE, Johansen E, Khamsi F: Cushing's syndrome associated with ectopic corticotropin production and small-cell lung cancer. J Clin Oncol 1992;10: 21–27.

13 Orth DN: Cushing's syndrome. N Engl J Med 1995;332:791–803.

14 Delisle L, Boyer MJ, Warr D, Killinger D, Payne D, Yeoh JL, Feld R: Ectopic corticotropin syndrome and small-cell carcinoma of the lung: clinical features, outcome, and complications. Arch Intern Med 1993;153:746–752.

15 Trainer PJ, Besser GM: St. Bartholomew's Hospital: The Bart's Endocrine Protocols. Edinburgh, Churchill-Livingstone, 1995.

16 Howlett TA, Drury PL, Perry L, Doniach I, Rees LH, Besser GM: Diagnosis and management of ACTH-dependent Cushing's syndrome: comparison of the features in ectopic and pituitary ACTH production. Clin Endocrinol (Oxf) 1986;24:699–713.

17 Sarlis NJ, Chanock SJ, Nieman LK: Cortisolemic indices predict severe infections in Cushing syndrome due to ectopic production of adrenocorticotropin. J Clin Endocrinol Metab 2000;85:42–47.

18 Putignano P, Kaltsas GA, Korbonits M, Jenkins PJ, Monson JP, Besser GM, Grossman AB: Alterations in serum protein levels in patients with Cushing's syndrome before and after successful treatment. J Clin Endocrinol Metab 2000;85:3309–3312.

19 Newell-Price J, Grossman A: Diagnosis and management of Cushing's syndrome. Lancet 1999;353:2087–2088.

20 Tabarin A, Corcuff JB, Rashedi M, Navarranne A, Ducassou D, Roger P: Comparative value of plasma ACTH and beta-endorphin measurement with three different commercial kits for the etiological diagnosis of ACTH-dependent Cushing's syndrome. Acta Endocrinol (Copenh) 1992;126:308–314.

21 Raffin-Sanson ML, Massias JF, Dumont C, Raux-Demay MC, Proeschel MF, Luton JP, Bertagna X: High plasma proopiomelanocortin in aggressive adrenocorticotropin-secreting tumors. J Clin Endocrinol Metab 1996;81:4272–4277.

22 Findling JW, Raff H: Diagnosis and differential diagnosis of Cushing's syndrome. Endocrinol Metab Clin N Am 2001;30:729–747.
23 Kaye TB, Crapo L: The Cushing syndrome: an update on diagnostic tests. Ann Intern Med 1990;112: 434–444.
24 Nieman LK, Oldfield EH, Wesley R, Chrousos GP, Loriaux DL, Cutler GB Jr: A simplified morning ovine corticotropin-releasing hormone stimulation test for the differential diagnosis of adrenocorticotropin-dependent Cushing's syndrome. J Clin Endocrinol Metab 1993;77: 1308–1312.
25 Terzolo M, Reimondo G, Ali A, Borretta G, Cesario F, Pia A, Paccotti P, Angeli A: The limited value of the desmopressin test in the diagnostic approach to Cushing's syndrome. Clin Endocrinol (Oxf) 2001;54:609–616.
26 Newell-Price J, Morris DG, Drake WM, Korbonits M, Monson JP, Besser GM, Grossman AB: Optimal response criteria for the human CRH test in the differential diagnosis of ACTH-dependent Cushing's syndrome. J Clin Endocrinol Metab 2002;87:1640–1645.
27 Reimondo G, Paccotti P, Minetto M, Termine A, Stura G, Bergui M, Angeli A, Terzolo M: The corticotrophin-releasing hormone test is the most reliable noninvasive method to differentiate pituitary from ectopic ACTH secretion in Cushing's syndrome. Clin Endocrinol (Oxf) 2003;58: 718–724.
28 Aron DC, Raff H, Findling JW: Effectiveness versus efficacy: the limited value in clinical practice of high dose dexamethasone suppression testing in the differential diagnosis of adrenocorticotropin-dependent Cushing's syndrome. J Clin Endocrinol Metab 1997;82:1780–1785.
29 Nieman LK, Chrousos GP, Oldfield EH, Avgerinos PC, Cutler GB Jr, Loriaux DL: The ovine corticotropin-releasing hormone stimulation test and the dexamethasone suppression test in the differential diagnosis of Cushing's syndrome. Ann Intern Med 1986;105:862–867.
30 Oldfield EH, Doppman JL, Nieman LK, Chrousos GP, Miller DL, Katz DA, Cutler GB Jr, Loriaux DL: Petrosal sinus sampling with and without corticotropin-releasing hormone for the differential diagnosis of Cushing's syndrome. N Engl J Med 1991;325:897–905.
31 Swearingen B, Katznelson L, Miller K, Grinspoon S, Waltman A, Dorer DJ, Klibanski A, Biller BM: Diagnostic errors after inferior petrosal sinus sampling. J Clin Endocrinol Metab 2004;89:3752–3763.
32 Yamamoto Y, Davis DH, Nippoldt TB, Young WF Jr, Huston J III, Parisi JE: False-positive inferior petrosal sinus sampling in the diagnosis of Cushing's disease. Report of two cases. J Neurosurg 1995;83:1087–1091.
33 Zarrilli L, Colao A, Merola B, La Tessa G, Spaziante R, Tripodi FS, Di Sarno A, Marzano LA, Lombardi G: Corticotropin-releasing hormone test: improvement of the diagnostic accuracy of simultaneous and bilateral inferior petrosal sinus sampling in patients with Cushing syndrome. World J Surg 1995;19:150–153.
34 McNally PG, Bolia A, Absalom SR, Falconer-Smith J, Howlett TA: Preliminary observations using endocrine markers of pituitary venous dilution during bilateral simultaneous inferior petrosal sinus catheterization in Cushing's syndrome: is combined CRF and TRH stimulation of value? Clin Endocrinol (Oxf) 1993;39:681–686.
35 Bonelli FS, Huston J III, Carpenter PC, Erickson D, Young WF Jr, Meyer FB: Adrenocorticotropic hormone-dependent Cushing's syndrome: sensitivity and specificity of inferior petrosal sinus sampling. Am J Neuroradiol 2000;21:690–696.
36 Colao A, Faggiano A, Pivonello R, Giraldi FP, Cavagnini F, Lombardi G: Inferior petrosal sinus sampling in the differential diagnosis of Cushing's syndrome: results of an Italian multicenter study. Eur J Endocrinol 2001;144:499–507.
37 Salenave S, Gatta B, Pecheur S, San Galli F, Visot A, Lasjaunias P, Roger P, Berge J, Young J, Tabarin A, Chanson P: Pituitary magnetic resonance imaging findings do not influence surgical outcome in adrenocorticotropin-secreting microadenomas. J Clin Endocrinol Metab 2004;89: 3371–3376.
38 Van Sickle DG: Carcinoid tumors. Analysis of 61 cases, including 11 cases of carcinoid syndrome. Cleve Clin Q 1972;39:79–86.
39 Shrager JB, Wright CD, Wain JC, Torchiana DF, Grillo HC, Mathisen DJ: Bronchopulmonary carcinoid tumors associated with Cushing's syndrome: a more aggressive variant of typical carcinoid. J Thorac Cardiovasc Surg 1997;114:367–375.

40 Torpy DJ, Chen CC, Mullen N, Doppman JL, Carrasquillo JA, Chrousos GP, Nieman LK: Lack of utility of (111)In-pentetreotide scintigraphy in localizing ectopic ACTH producing tumors: follow-up of 18 patients. J Clin Endocrinol Metab 1999;84:1186–1192.
41 De Herder WW, Lamberts SW: Octapeptide somatostatin-analogue therapy of Cushing's syndrome. Postgrad Med J 1999;75:65–66.
42 Tabarin A, Valli N, Chanson P, Bachelot Y, Rohmer V, Bex-Bachellerie V, Catargi B, Roger P, Laurent F: Usefulness of somatostatin receptor scintigraphy in patients with occult ectopic adrenocorticotropin syndrome. J Clin Endocrinol Metab 1999;84:1193–1202.
43 Tsagarakis S, Christoforaki M, Giannopoulou H, Rondogianni F, Housianakou I, Malagari C, Rontogianni D, Bellenis I, Thalassinos N: A reappraisal of the utility of somatostatin receptor scintigraphy in patients with ectopic adrenocorticotropin Cushing's syndrome. J Clin Endocrinol Metab 2003;88:4754–4758.
44 Fanti S, Farsad M, Battista G, Monetti F, Montini GC, Chiti A, Savelli G, Petrella F, Bini A, Nanni C, Romeo A, Franchi R, Bombardieri E, Canini R, Monetti N: Somatostatin receptor scintigraphy for bronchial carcinoid follow-up. Clin Nucl Med 2003;28:548–552.
45 Doppman JL, Nieman L, Miller DL, Pass HI, Chang R, Cutler GB Jr, Schaaf M, Chrousos GP, Norton JA, Ziessman HA: Ectopic adrenocorticotropic hormone syndrome: localization studies in 28 patients. Radiology 1989;172:115–124.
46 Doppman JL, Pass HI, Nieman LK, Miller DL, Chang R, Cutler GB Jr, Chrousos GP, Jaffe GS, Norton JA: Corticotropin-secreting carcinoid tumors of the thymus: diagnostic unreliability of thymic venous sampling. Radiology 1992;184:71–74.
47 White A, Gibson S: ACTH precursors: biological significance and clinical relevance. Clin Endocrinol (Oxf) 1998;48:251–255.
48 Aniszewski JP, Young WF Jr, Thompson GB, Grant CS, van Heerden JA: Cushing syndrome due to ectopic adrenocorticotropic hormone secretion. World J Surg 2001;25:934–940.
49 Drake WM, Perry LA, Hinds CJ, Lowe DG, Reznek RH, Besser GM: Emergency and prolonged use of intravenous etomidate to control hypercortisolemia in a patient with Cushing's syndrome and peritonitis. J Clin Endocrinol Metab 1998;83:3542–3544.
50 Krakoff J, Koch CA, Calis KA, Alexander RH, Nieman LK: Use of a parenteral propylene glycol-containing etomidate preparation for the long-term management of ectopic Cushing's syndrome. J Clin Endocrinol Metab 2001;86:4104–4108.
51 Greening JE, Brain CE, Perry LA, Mushtaq I, Sales MJ, Grossman AB, Savage MO: Efficient short-term control of hypercortisolaemia by low-dose etomidate in severe paediatric Cushing's disease. Horm Res 2005;64:140–143.

Prof. Ashley B. Grossman
St Bartholomew's Hospital
London, EC1A 7BE (UK)
Tel. +44 207 601 8343, Fax +44 207 601 8505, E-Mail a.b.grossman@qmul.ac.uk

Arzt E, Bronstein M, Guitelman M (eds): Pituitary Today: Molecular, Physiological and Clinical Aspects. Front Horm Res. Basel, Karger, 2006, vol 35, pp 157–168

Estrogens and Neuroendocrine Hypothalamic-Pituitary-Adrenal Axis Function

Alejandro F. De Nicola[a,b], *Flavia E. Saravia*[a,b], *Juan Beauquis*[a], *Luciana Pietranera*[a,b], *Monica G. Ferrini*[c]

[a]Laboratory of Neuroendocrine Biochemisty, Instituto de Biología y Medicina Experimental, Buenos Aires, [b]Department of Human Biochemistry, Faculty of Medicine, University of Buenos Aires, Buenos Aires, Argentina; [c]Research and Education Institute, Harbor-UCLA Medical Center, Urology, Torrance, Calif., USA

Abstract

The function of the HPA axis is subject to regulation by many factors, which achieve relevance under normal and pathological conditions. In the case of aging, this period of life is associated with disturbances of the HPA axis and signs of hippocampal vulnerability. We examined 20-month-old male rats, in which abnormalities of the HPA axis included altered response to stress, reduced effectiveness of the steroid negative feedback and low expression of hippocampal glucocorticoid receptors (GR). Estrogen treatment of aging rats normalized the response to stress, restored the dexamethasone inhibition of the stress response and increased GR density in defined hippocampal areas. Although estrogens could influence the hippocampus of aging animals directly, their effects could be also mediated by estrogen-sensitive forebrain cholinergic neurons projecting to the hippocampus. Additionally, estrogens normalized the deficient granule cell proliferation that aging mice present in the dentate gyrus, and attenuated several markers of hippocampal aging, such as astrocytosis, high lipofucsin content and neuronal loss in the hilus of the dentate gyrus. These effects may be important for the regulation of the HPA axis, in the context that hippocampal function as a whole was normalized by estrogen action. Therefore, estrogens are powerful neuroprotectants in cases of hippocampal dysfunction, and as part of this effect, they contribute to stabilize the function of the HPA axis.

Physiological Aspects of HPA Axis Function

The adrenal cortex is only in part an autonomous endocrine gland, because its function is mainly dependent on pituitary ACTH. In turn, synthesis and secretion of pituitary ACTH is under the regulatory control of neuropeptides synthesized in the paraventricular nucleus (PVN) of the hypothalamus, i.e. the corticotropin releasing hormone (CRH) and arginine vasopressin (AVP). Both peptides are cosynthesized in the parvocellular division of the PVN, whereas cells in the magnocellular division of the PVN synthesize AVP and oxytocin. Parvocellular CRH and AVP are secreted from axon terminals at the zona externa of the median eminence into long portal vessels bathing the anterior pituitary. Magnocellular-derived AVP may also reach the anterior pituitary. In this case, AVP is released en passant from axons of the hypothalamic-neurohypophysial system at the median eminence which gain access to the long portal vessels [1]. During stress, it is likely that AVP acting on the anterior pituitary originates from both the parvocellular and magnocellular divisions of the PVN [2]. CRH and AVP regulate the secretory activity of pituitary corticotropes in a coordinate manner, whereas CRH – the main ACTH secretagogue – also increases ACTH synthesis from the precursor molecule proopiomelanocortin (POMC) [3].

CRH secretion is driven in a circadian manner, and both CRH and AVP are stimulated by stress and factors acting on the CNS including chemicals, change of environmental conditions, fear and anxiety, hemorrhage, hypoxia, hypoglycemia, infections, etc. [3]. These factors induce the release of excitatory neurotransmitters (serotonin, acetylcholine and catecholamines), cytokines, angiotensin II and several neuropeptides, or attenuate inhibitory neurotransmission due to GABA and opiates. Thus, control of CRH and AVP synthesis and release from the PVN can be both stimulatory or inhibitory. Three main neuroanatomical pathways regulate the activity of the PVN [4]. The first originates in the suprachiasmatic nucleus and is responsible for the circadian variation of ACTH and circulating glucocorticoids. The second is an stimulatory adrenergic projection to the PVN from the locus coeruleus and the nucleus of the tractus solitarious found in the brain stem. The third is inhibitory, represented by a GABAergic pathway originated in the hippocampus which ends in the PVN via the stria terminalis. The last constitutes the main pathway linking the hippocampus and the PVN, due to the lack of direct innervation between these structures. As discussed later, it plays an important role in the hippocampal inhibitory control of the HPA axis.

Pituitary ACTH derives from POMC processing. POMC breakdown products are the amino-terminal peptide 1–76, the joining peptide 1–30, ACTH 1–39 and β-lipotropin 1–89 [3]. This processing is increased by CRH, which binds to pituitary CRH1 receptors coupled to the cAMP/protein kinase A system. Once formed, ACTH secretion is stimulated by CRH and AVP. Whereas CRH is involved in the normal ACTH circadian drive, stimulation by AVP preferentially occurs after some forms of stress and in stress-related diseases [1, 2].

AVP binds to V1b (also called V3) receptors in the anterior pituitary, and stimulates ACTH secretion after activation of the phospholipase C-mediated hydrolysis of phosphatidylinositol 4,5-biphosphate [5]. Thus, CRH and AVP act in a complementary manner, using separate signaling cascades, to increase the secretory activity of pituitary corticotropes.

Steroid Inhibitory Feedback Control of the HPA Axis under Normal and Abnormal Conditions

Natural glucocorticoids – cortisol in humans and corticosterone (CORT) in rodents – are secreted from the fasciculata zone of the adrenal cortex under the influence of ACTH. Glucocorticoids bind to the glucocorticoid receptor (GR) and the mineralocorticoid receptor (MR), a family of nuclear transcription factors which are expressed in a region-specific manner in the CNS and anterior pituitary [4]. GR and MR can dimerize, heterodimerize or act as monomers, originating alternative mechanisms for the control of the expression or repression of genes and for the modulation of non-genomic events taking place in the cytoplasm [4]. Three main anatomical sites are involved in steroid inhibitory feedback of the HPA axis, namely the anterior pituitary corticotrope, the hypothalamic PVN and the hippocampus. The neuropeptide-producing cells of the parvocellular division of the PVN and pituitary corticotropes express GR, a low affinity receptor that mediates steroid effects under conditions of high circulating hormone levels, i.e. during stress and at the peak of the circadian rhythm and is also involved in the termination of stress responses [6]. Instead, the hippocampus is enriched in GR and MR, the latter being a promiscuous receptor that binds with high affinity both mineralocorticoids and glucocorticoids [6, 7]. However, hippocampal MR prefers CORT, which circulates in high amounts, rather than aldosterone, which circulates in minor amounts. This peculiarity is due to the low expression of the type 2 isoform of the enzyme 11β-hydroxysteroid dehydrogenase (11-HSD2) in the hippocampus. 11-HSD2 oxidizes cortisol and CORT to inactive metabolites, enabling the access of aldosterone to MR. This mechanism occurs in the circumventricular organs or the periphery [8]. Preferential occupancy of hippocampal MR by glucocorticoids, caused by the low activity of 11-HSD2, may be responsible for the fine inhibitory control of the HPA axis. Glucocorticoid binding to MR is also favored by reactive oxygen species (ROS) produced under pathological conditions. MR may thus be converted into a 'death receptor', because it can originate vascular damage in the brain and peripheral organs [9].

Binding of glucocorticoids to GR in the hippocampus, PVN and pituitary corticotropes regulates the function of the HPA axis under conditions of stress or peak circadian steroid levels. However, steroid feedback inhibition at the pituitary level may be less significant in a normal steroid environment, because the plasma membrane of corticotropes contains a CBG-line protein that prevents the access of natural glucocorticoids to intracellular GR [10]. In contrast, the synthetic glucocorticoid dexamethasone (DEX) lacks affinity for CBG and is easily bound by anterior pituitary GR.

There are numerous situations commonly associated with disturbances in the function of the HPA axis. Most of them are stress-related diseases, including pathological aging, diabetes mellitus, depression, post-traumatic stress disorder (PTSD) and neurodegeneration of the Alzheimer's type. These diseases, with the exception of PTSD, present increased levels of circulating glucocorticoids, a sustained response to stress, flattening of the diurnal rhythm of plasma corticosteroids and ACTH, decreased glucocorticoid negative feedback and hippocampal neuropathology [11, 12]. Signs of hippocampal vulnerability include GR downregulation, atrophy of apical dendrites of pyramidal neurons in the CA3 region, astrogliosis, reduced neurogenesis in the dentate gyrus and possible neuronal loss [13, 14]. Inappropriate activation of GR by high levels of circulating glucocorticoids induces receptor downregulation, because GR acts as a natural repressor acting at the promoter of the GR gene.

In aging rodents, basal levels of circulating glucocorticoids are normal or higher than normal but stress-induced levels of glucocorticoids persist longer than normal [11, 12]. The amount of corticosteroid receptors present in the hippocampus of old rats is controversial, although most authors, including ourselves, have reported decreased hippocampal GR and MR, reviewed in [15]. In cases of GR deficiency, it is hypothesized that diminished steroid feedback enhances the deleterious effects of excess levels of circulating glucocorticoids. This possibility is based on evidence that hippocampal neurons are highly vulnerable to increased glucocorticoid action [4, 11]. Hippocampal damage releases the inhibitory GABAergic tone exerted upon the hypothalamus, resulting in increased production of hypothalamic neuropeptides. For example, hippocampectomized animals show increased synthesis of CRH and AVP mRNA and decreased sensitivity to steroid feedback [16, 17]. The effect of hippocampectomy is reproduced in other animal models. For example, GR heterozygous mutant mice (GR+/−) show a disinhibited HPA system and a pathological DEX test. Although to a lesser extent, GR deficiency leads to similar alterations in aging and experimental diabetes mellitus. Thus, elevated levels of circulating CORT, hyperresponse to stress, flattening of circadian rhythms, increased synthesis of hypothalamic neuropeptides, down-regulation of GR and lack of the inhibitory feedback effect of glucocorticoid are typical findings in rodents with type I diabetes induced by streptozotocin or in genetic forms of the disease, reviewed in [18]. Both aging animals and streptozotocin-diabetic animals are useful models to analyze plastic changes of the HPA axis, mainly the possibility that changes are not permanent but reversible by factors modifying the function of the HPA axis, including estrogens.

Neuroendocrine Effects of Estrogens

In contrast to the pathological consequences that excess levels of glucocorticoids produce in the hippocampus, estrogens qualified as brain 'neuroprotectans'. Some estrogen effects could be mediated by genomic mechanisms. Estrogen binding has been reported in the hippocampal pyramidal cells and the hilus of the dentate gyrus. Of the two isoforms of the estrogen receptor, ERα and ERβ, the newly described β isoform is abundantly expressed in hippocampus, whereas ERα is found in CA1 interneurons and a subset of pyramidal and granule cells [19]. McEwen et al. [20] found that estrogens increase dendritic spine formation and synaptic density in CA1 pyramidal cells, an effect probably mediated by ER. In hippocampal neurons in culture, estrogens protect against glutamate toxicity, glucose deprivation, FeSO toxicity and amyloid-β peptide toxicity, the hallmark of Alzheimer's disease, reviewed in Goodman et al. [21]. Interestingly, CORT exacerbated oxidative stress caused by these agents. Estrogens play an important role in the hippocampus, related to learning and memory. One way estrogens modulate these processes is by enhancement of neurogenesis – the proliferation, migration and differentiation of new neurons – in the dentate gyrus of adult animals. In this region, and in the subventricular zone, neurogenesis continues into adulthood, although it declines in aging animals [22]. Both ERα and ERβ mRNA are found in ≈80% of proliferating cells of the dentate gyrus labelled with the Ki67 antibody, and in an important proportion of cells showing a more mature phenotype [23]. We favor the view that generation of new neurons in the dentate gyrus under the influence of estrogens may be also important for the maintenance of hippocampal function and for adequate control of the HPA axis.

Previous work demonstrated that adrenal status is highly dependent on estrogens. In rodents, there are sex differences in circulating glucocorticoid levels as well as in adrenal size, which is higher in female rats. Estrogens also increase CBG, the main glucocorticoid transport protein in plasma. Acting upon the anterior pituitary and hypothalamus, estrogens play a facilitatory role on ACTH and CORT release during stress. Also, GR are targets of estrogens, as shown by increases in GR mRNA in the amygdala and hypothalamus of estrogen-treated animals [24]. We have previously reported that prolonged estrogen treatment increases GR binding and immunoreactivity in several brain regions, while it also abolishes GR down-regulation in rats exposed to chronic glucocorticoid treatment [25, 26]. Other laboratories have also shown that estrogens and phytoestrogens enhance GR abundance in hippocampus and regulate the HPA axis [27]. Working on the anterior pituitary and the PVN, estrogens down-regulate transcription of the CRH and AVP genes in the PVN and the POMC gene in the corticotrope [29]. Both ER isoforms seem to be involved, since ERβ critically modulates the HPA axis response to stress, whereas ERα may directly influence CRH neurons. As already known, AVP mRNA expression is reduced after DEX treatment. Because estrogen treatment enhances the DEX effect on AVP mRNA, the effect may be due to increased GR levels present in estrogen-treated animals. Altogether, our studies raise the possibility that positive modulation of GR by estrogens could reinforce the glucocorticoid feedback mechanism. In human studies, stress-induced glucocorticoid elevations in postmenopausal women are blunted by estrogen replacement.

Estrogens and HPA Axis in Aging

In our laboratory, we aimed to determine whether estrogen treatment modifies some neuroendocrine parameters altered in aging rats. We selected for this study the response to stress, feedback inhibition and the expression of GR in hippocampus and hypothalamus. The use of male animals was deliberately chosen because of the high incidence of pituitary tumors in old female rodents [see chapter by Goya et al, this vol] which by itself may disrupt glucocorticoid feedback mechanisms. Additionally, most research on estrogen effects was carried out in females, and we wished to test the sensitivity of old male animals to female steroids. In our studies, young 3- to 4-month-old and old 18- to 20-month-old male Sprague-Dawley rats were implanted with a single pellet of estradiol benzoate weighing 14 mg for 6 weeks [15]. This procedure increased plasma estradiol levels by more than 10-fold (≈300–400 pg/ml). These levels were clearly pharmacological, although it is possible that local estrogen synthesis in the brain due to aromatase activity outnumbers circulating estradiol levels. The biological response to estrogens was corroborated by significant increments in pituitary weight and plasma prolactin.

Basal levels of CORT did not show an age effect. Mean levels were around 5–10 μg/ml, but exposure to ether vapors for 1 min produced a 5- to 7-fold stimulation of plasma CORT levels in all groups at 15 min poststress, without significant differences between young and old rats, either untreated or receiving estradiol. In young animals both untreated or estradiol-treated, plasma CORT returned to near basal values 2 h post-stress. However, old rats did not switch off the CORT response to ether stress, resulting in plasma CORT levels at 2 h similar to those at 15 or 60 min. In contrast to young rats in which estrogen did not influence the stress response, estradiol treatment of old rats restored the CORT response to ether stress at 120 min to levels comparable to those of young or young + estradiol-treated rats. Considering that estradiol treatment normalized the stress response of old rats, these animals were subsequently evaluated regarding the effects of DEX inhibition on plasma CORT. Old animals partially 'escaped' from DEX inhibition and significantly raised their plasma CORT levels 15 min following ether stress. In contrast, estradiol treatment of old animals seemed to sensitize glucocorticoid suppression of the HPA axis, since DEX completely blunted the response to ether stress of the estradiol-treated old rats.

Changes in dynamics of plasma CORT suggested that disruption of feedback mechanisms in old rats could be normalized by estradiol treatment. To study whether these changes correlated with changes in GR levels, quantitative densitometry of GR immunoreaction was carried out in the PVN and five hippocampal subfields: CA1, CA2, CA3, CA4 and subiculum. In the CA1 and CA2 subfields and the subiculum, untreated old rats showed a reduction in GR with respect to young and young + estradiol. Estrogen treatment of old rats increased GR at levels shown by young and young + estradiol. In contrast to the aging and estradiol effects in the CA1, CA2 subfields and subiculum, no changes were registered for aging or estradiol in the CA3 and CA4 subfields or the hypothalamic PVN. Therefore, estradiol produced a region-specific modulation of GR immunoreactivity when given to old, but not young animals. These data are in agreement with Pfeiffer et al. [24] and Lephart et al. [27] who demonstrated that estrogen and estrogenic-like compounds increase hippocampal GR. In conclusion, 6 weeks of estrogen exposure normalized the response to stress, restored feedback inhibition and increased hippocampal GR immunoreactivity in old rats. It is likely that the abnormal stress response reflected a dysfunction of the negative feedback effect of glucocorticoids. Feedback mechanisms are usually mediated by steroid receptors localized in the CNS and pituitary, although lesion and aging studies point to the hippocampus as the area with greater receptor plasticity. This is because lesions of the hippocampus associate with abnormal shut-off of the HPA stress response [16, 17], whereas reports in old animals have demonstrated a delayed termination of stress responses coincident with reduced hippocampal MR and GR levels.

Estrogen Effects on the Hippocampus May Be Direct or Mediated by the Forebrain Cholinergic System

Detection of ER isoforms, of membrane-associated receptors, the modulation of ion channels and cytoplasmic signaling cascades [29] suggest that sex hormone effects in aging animals may directly impinge on the hippocampus.

However, indirect effects on the neuroendocrine hippocampus due to estrogen primary effects on hormone-sensitive forebrain structures are also likely. Two forebrain regions, the medial septum (MS) and the vertical limb of the diagonal band of Brocca (VDB), innervate the hippocampal subfields CA2 and CA3 and the dentate gyrus. These projections are cholinergic, comprising about 75% of the acetylcholine delivered to the hippocampus. Cholinergic cells are positive for both ERα and ERβ mRNA and are hormone sensitive, as shown by the ability of estrogens to induce the expression of the enzymes choline-acetyltransferase (ChAT) and cholinesterase and by pronounced hormonal effects on axonal sprouting in the forebrain [29].

Thus, we evaluated the possibility that modulation by estrogens of the response to stress and prevention of the neurotoxic consequences of increased HPA activity in the aged group could be mediated by the cholinergic system. To this end, we studied ChAT immunoreactivity of neurons and cell processes in the MS and VDB of untreated young, young + estradiol, untreated old and old + estradiol-treated male rats. Animal treatment was identical to that used in experiments designed to study estrogen effects on the stress response and immunoreactive GR levels of aging animals [15]. Aging animals showed a reduction in the number and length of ChAT-immunoreactive cell processes, but not of immunopositive neurons in MS and VDB. Estradiol reverted both parameters in old rats to levels of young animals [30]. In a parallel study, we quantitatively assessed the levels of the growth-associated protein (GAP43) mRNA, a marker for neurite outgrowth, using in situ hybridization techniques. Although basal levels of GAP43 mRNA were similar in old and young animals, estradiol treatment upregulated this messenger in MS and VDB of old animals. In this study, we concluded that estrogens enhanced the plasticity of cholinergic, GAP43 expressing neurons of the basal forebrain. These and previous data, therefore, sustain that estrogen effects on the hippocampus, including the fine control of the HPA axis of aging animals, may depend on two separate estrogen-responsive mechanisms. A first mechanism would be directly exerted on the hippocampus, whereas a second one could be mediated by the forebrain cholinergic system.

Estrogen Effects on Hippocampal Neurogenesis: A New Window for HPA Axis Regulation?

After maturation, cells originating in the subgranular zone of the dentate gyrus become mature granule cells, which extend their axons – the so-called mossy fibers – towards the hilus, while a long axonal branch make synapses with dendrites of the CA3 pyramidal neurons. Axons of CA3 project to the CA1 via branches extending from the Schaffer collaterals. Therefore, granule neurons make contact with the CA1 and CA3 regions, becoming integrated within the hippocampal circuitry. At one time point or another of their existence, proliferating and maturing granular cells express MR and GR [32], making them susceptible to changes in circulating adrenocorticoid hormones. Thus, adrenalectomy increases neurogenesis whereas high levels of circulating glucocorticoids inhibits it [22]. The neural progenitors of the subgranular zone of the dentate gyrus are also targets of estrogens, since 80% of these cells express the α and β isoforms of ER [23]. Estrogens enhance neurogenesis, in contrast to the deleterious effects of glucocorticoids. This effect resembles the opposing effects of glucocorticoids and estrogens on GR. Thus, excess glucocorticoids – as seen in aging animals – downregulate GR, whereas estrogen treatment of aging animals increases GR in hippocampus.

To study the influence of estrogens on brain neurogenesis in aging rodents, we examined 9–11 month old male C57BL/6 mice, in which we implanted a single estradiol pellet weighing 150 μg or a cholesterol vehicle pellet. Animals were injected i.p. with bromodeoxyuridine (BrdU), a modified nucleotide that incorporates into the DNA of dividing cells at the S phase. 4 and 2 h before killing, animals received BrdU and brains were analyzed for BrdU-specific staining. Untreated old mice showed a pronounced reduction of BrdU incorporation (about 250 cells per dentate gyrus) but this figure was increased by 2-fold after animals were exposed to estrogens. At high microscope magnification, estrogen-treated old mice showed clusters of proliferating cells, an image never seen in untreated old mice. The hippocampus of untreated old mice showed a remarkable aging effect when they were 9–11 months old. Thus, astrocytosis (an index of neuronal suffering), increased neuronal content of lipofuscine (an index of aging) and loss of hilar neurons (an index of neuronal damage) confirmed an aged-related neurodegeneration. Estrogen completely reverted these changes, because astrocyte number, lipofucsin content and hilar neuronal number were all recovered to levels of young mice. Therefore, estrogen behaved as a powerful neuroprotectant to the aging hippocampus, since increased neurogenesis under the influence of estrogens was accompanied by restoration of hippocampal function. Interestingly, both estrogen-enhanced neurogenesis and estrogen control of HPA activity and hippocampal GR required a permissive environment (i.e. aging), and did not occur in young animals. This parallel effect suggest that neurogenesis and HPA axis control may be somehow related.

Conclusions

Estrogens control reproductive and nonreproductive events in the brain, and also play a role in the function of the HPA axis. Concerning the last aspect, structures implicated in HPA axis control such as the hippocampus, hypothalamus and the anterior pituitary, are targets of estrogens. Our studies focused on aging rodents, which presented a marked hyperfunction of the HPA axis. We confirmed that old animals show abnormalities in the response to stress, altered feedback mechanism and low expression of hippocampal GR. Estrogen treatment of aging rats normalized the response to stress, restored the DEX inhibition of the stress response and increased GR density in defined hippocampal areas. Although estrogens could influence the hippocampus of aging animals directly, their effects could be also mediated by the forebrain cholinergic system, which is a known target of estrogens. We found that in old animals, estrogens restored to normal the deficient ChAT immunoreactivity and GAP43 mRNA in forebrain cholinergic nuclei projecting to the hippocampus. We also studied the estrogen effects on cell proliferation in the dentate gyrus of aging animals. The latter may be an additional control mechanism of the HPA axis, because newborn cells in the subgranular zone of the dentate gyrus differentiate, send projections to the pyramidal cells and integrate into the hippocampal circuitry. It is possible that in aging animals, estrogens modulated the neuroendocrine hippocampus by increasing cell proliferation in the dentate gyrus, and also by attenuation of several markers of aging, such as astrocytosis, high lipofucsin content and neuronal loss in the hilus of the dentate gyrus. In aging animals, hippocampal function and the steroid negative feedback may be re-established by estrogens, stabilizing the function of the HPA axis. Thus, estrogen regulation of the HPA axis remains an exciting area of research to further understand hormonal disbalances in the aging hippocampus.

References

1 Engelmann M, Landgraf R, Wotjak CT: The hypothalamic-neurohypophysial system regulates the hypothalamic-pituitary-adrenal axis under stress: an old concept revisited. Front Neuroendocrinol 2004;25:132–149.

2 Angulo JA, Ledoux M, McEwen BS: Genomic effects of cold and isolation stress on magnocellular vasopressin mRNA-containing cells in the hypothalamus of the rat. J Neurochem 1991;23: 2033–2038.

3 Wilson JD, Foster DW, Kronenberg HM, Larsen PR (eds): Williams Textbook of Endocrinology. Philadelphia, Saunders, 1998.

4 De Kloet ER, Joels M, Holsboer F: Stress and the brain: from adaptation to disease. Nat Rev Neurosci 2005;6:463–475.

5 Hurbin A, Boissin-Agasse L, Orcel H, Rabié A, Joux N, Desarrménien G, Richard P, Moos FC: The V1a and V1b, but not the V2, vasopressin receptor genes are expressed in the supraoptic nucleus of the rat hypothalamus, and the transcripts are essentially colocalized in the vasopressinergic magnocellular neurons. Endocrinology 1998;139:4701–4707.

6 Ratka A, Sutanto W, Bloemers M, De Kloet ER: On the role of brain mineralocorticoid (type I) and glucocorticoid (type II) receptors in neuroendocrine regulation. Neuroendocrinology 1989;50: 117–123.

7 Magariños AM, Ferrini M, De Nicola AF: Corticosteroid receptors and glucocorticoid content in microdissected brain regions: correlative aspects. Neuroendocrinology 1989;50:673–678.
8 Edwards CR, Stewart PM, Burt D, McIntyre MA, de Kloet ER, Brett L, Sutanto W, Monder C: Localisation of 11beta-hydroxysteroid dehydrogenase: tissue-specific protector of the mineralocorticoid receptor. Lancet 1988;ii:986–999.
9 Funder JW: Aldosterone, mineralocorticoid receptors and vascular inflammation. Mol Cell Endocrinol 2004;217:263–269.
10 Perrot-Applanat M, Racadot O, Milgrom E: Specific localization of plasma corticosteroid-binding globulin in pituitary corticotrophs. Endocrinology 1984;115:559–569.
11 Sapolsky R, Krey LC, McEwen BS: The neuroendocrinology of stress and aging: the glucocorticoid cascade hypothesis. Endocr Rev 1986;7:284–304.
12 Van Eekelen JAM, Rots NY, Sutanto W, de Kloet ER: The effect of aging on stress responsiveness and central corticosteroid receptors in the Brown Norway rat. Neurobiol Aging 1991;13:159–170.
13 Tornello S, Orti E, De Nicola AF, Rainbow TC, McEwen BS: Regulation of glucocorticoid receptors in brain by CORT treatment of adrenalectomized rats. Neuroendocrinology 1982;35:411–417.
14 Watanabe Y, Gould E, McEwen BS: Stress induces atrophy of apical dendrites of hippocampal CA3 pyramidal neurons. Brain Res 1992;588:341–345.
15 Ferrini M, Piroli G, Frontera M, Falbo A, Lima A, De Nicola AF: Estrogens normalize the hypothalamic-pituitary-adrenal axis response to stress and increase glucocorticoid receptor immunoreactivity in hippocampus of aging male rats. Neuroendocrinology 1999;69:129–137.
16 Herman JP, Schafer MK, Young EA, Thompson R, Doiuglass J, Akil H, Watson SJ: Evidence for hippocampal regulation of neuroendocrine neurones of the hypothalamic-pituitary-adrenocortical axis. J Neurosci 1989;9:3071–3082.
17 Magariños AM, Somoza G, De Nicola AF: Glucocorticoid negative feedback and glucocorticoid receptors after hippocampectomy in rats. Horm Metab Res 1987;19:105–109.
18 Revsin Y, Saravia F, Roig P, Lima A, de Kloet ER, Homo-Delarche F, De Nicola AF: Neuronal and astroglial alterations in the hippocampus of a mouse model for type 1 diabetes. Brain Res 2005;1038:22–31.
19 Shughrue PJ, Lane MV, Merchentaler I: Comparative distribution of estrogen receptor α and β mRNA in the rat central nervous system. J Comp Neurol 1997;388:507–525.
20 Gould E, Wooley CS, Frankfurt M, McEwen BS: Gonadal steroids regulate dendritic spine density in hippocampal pyramidal cells in adulthood. J Neurosci 1990;10:1286–1291.
21 Goodman Y, Bruce AJ, Cheng B, Mattson MP: Estrogens attenuate and CORT exacerbates excitotoxicity, oxidative injury, and amyloid-peptide toxicity in hippocampal neurons. J Neurochem 1996;66:1836–1844.
22 Cameron H, McKay RDG: Restoring production of hippocampal neurons in old age. Nature Neurosci 1999;2:894–897.
23 Isgor C, Watson SJ: Estrogen receptor α and β mRNA expressions by proliferating and differentiating cells in the adult rat dentate gyrus and subventricular zone. Neuroscience 2005;134: 847–856.
24 Pfeiffer A, Lapointe B, Barden N: Hormonal regulation of type II glucocorticoid receptor messenger ribonucleic acid in rat brain. Endocrinology 1991;129:2166–2174.
25 Ferrini M, De Nicola AF: Estrogens up-regulate type I and type II glucocorticoid receptors in brain regions from ovariectomized rats. Life Sci 1991;48:2593–2601.
26 Ferrini M, Lima A, De Nicola AF: Estradiol abolishes down-regulation of glucocorticoid receptors in brain. Life Sci 1995;57:2403–2412.
27 Lephart ED, Galindo E, Bu LH: Stress (hypothalamic-pituitary-adrenal axis) and pain response in male rats exposed lifelong to high vs. low phytoestrogen diets. Neurosci Lett 2003;342:65–68.
28 Isgor C, Cecchi M, Kabbai M, Akil H, Watson SJ: Estrogen receptor beta in the paraventricular nucleus of hypothalamus regulates the neuroendocrine response to stress and is regulated by CORT. Neuroscience 2003;121:837–845.
29 Nomura M, McKenna E, Korach KS, Pfaff DW, Ogawa S: Estrogen receptor-beta regulates transcript levels for oxytocin and arginine vasopressin in the hypothalamic paraventricular nucleus of male mice. Brain Res Mol Brain Res 2002;30:109:84–94.
30 McEwen BS: Estrogen actions throughout the brain. Rec Progr Horm Res 2002;57:357–384.

31 Ferrini M, Bisagno V, Piroli G, Grillo C, Gonzalez Deniselle MC, De Nicola AF: Effects of estrogens on choline-acetyltransferase immunoreactivity and GAP43 mRENA in the forebrain of young and aging male rats. Cell Mol Neurobiol 2002;22:289–301.
32 Wong EYH, Herbert J: Roles of mineralocorticoid and glucocorticoid receptors in the regulation of progenitor proliferation in the adult hippocampus. Eur J Neurosci 2005;22:785–792.

Dr. Alejandro F. De Nicola
Instituto de Biología y Medicina Experimental
Obligado 2490
1428 Buenos Aires (Argentina)
Tel. +54 11 4783 2869, Fax +54 11 4786 2564, E-Mail denicola@dna.uba.ar

Arzt E, Bronstein M, Guitelman M (eds): Pituitary Today: Molecular, Physiological and Clinical Aspects. Front Horm Res. Basel, Karger, 2006, vol 35, pp 169–178

New Aspects in the Diagnosis and Treatment of Cushing Disease

M. Labeur[a], *M. Theodoropoulou*[a], *C. Sievers*[a], *M. Paez-Pereda*[b], *V. Castillo*[c], *E. Arzt*[d], *G.K. Stalla*[a]

[a]Max-Planck-Institute of Psychiatry, [b]Affectis Pharmaceuticals, Munich, Germany; [c]Facultad de Ciencias Veterinarias, UBA, [d]Laboratorio de Fisiología y Biología Molecular, FCEN, UBA and CONICET, Buenos Aires, Argentina

Abstract

Cushing disease which is caused by the excessive production of ACTH is a rare and complex endocrine disorder that still represents a major challenge for the physician in terms of accurate diagnosis and efficient treatment. Diagnosing Cushing syndrome and its etiology is an elaborate procedure and no single test is sensitive and specific enough to provide sufficient accuracy. Therefore, an ordered cascade of tests is necessary recommended by a consensus statement in 2002. The proposed diagnostic algorithm will be summarized in the following section. In the absence of efficient drug therapy, transsphenoidal resection of the pituitary adenoma is the treatment of choice for the reduction of ACTH secretion. However, not all patients can be cured by surgery. In the present article, we examine recent studies that have investigated the therapeutic potential of new generations of drugs for the treatment of Cushing disease such as cabergoline and SOM230. The role of nuclear receptors: retinoic acid receptors and peroxisome proliferator-activated receptor-γ as new approaches for treating pituitary tumors is also discussed.

Diagnosis of Cushing Disease

In October 2002, the international consensus conference with 50 leading endocrinologists in Ancona, Italy (organized by G. Arnaldi and M. Boscaro) set the basis for the current 'state-of-the-art' diagnostic procedure in the management of Cushing syndrome (CS) (fig. 1) [1]. The following section gives an overview of the main recommendations.

Regardless of etiology, all cases of endogenous CS are due to the increased production of cortisol by the adrenal gland which leads to the specific symptoms.

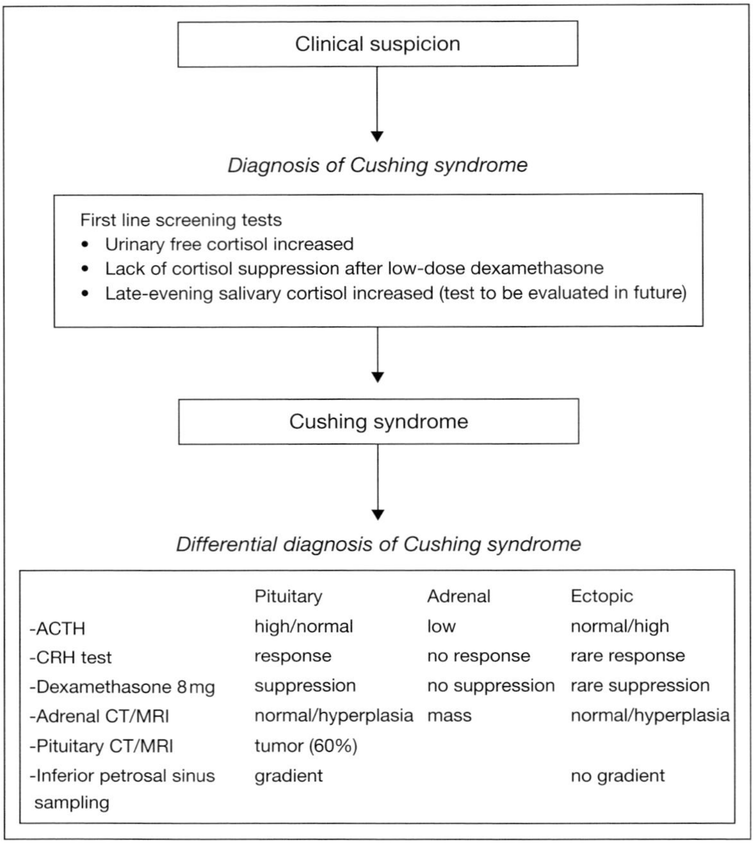

Fig. 1. Diagnostic algorithm according to the recommendations of Arnaldi et al. [1].

All diagnostic tests are based on the principle of inappropriate cortisol secretion with loss of negative feedback inhibition.

A requisite to start any diagnostic procedure is a sufficient clinical suspicion. CS produces a myriad of disabling and sometimes life-threatening symptoms, including typical abnormal fat deposition around the neck, thinning of the skin, osteoporosis, insulin resistance, dyslipidemia, myopathy, amenorrhea and hypertension. Fatigue, irritation, anxiety and depression are also common clinical features in these patients [2].

Clinical manifestations can vary due to the duration and extent of hypercortisolism which spans from high oversecretion to barely detectable or cyclical excretion. Some patients may present with isolated symptoms and lack even the most common findings like hypertension or truncal obesity. Therefore, CS can easily be misdiagnosed.

Diagnosis of Cushing Syndrome

If CS is suspected, the 24-hour urinary free cortisol assay and the low-dose dexamethasone suppression test are the screening tests of choice.

The 24-hour urinary free cortisol tests are often positive in pseudo-Cushing states such as depression and alcoholism, but values four times higher than normal are considered diagnostic for CS. The test should be repeated up to three times in case of inconclusive results.

In the low-dose dexamethasone suppression test (dexamethasone 1 or 2 mg overnight), a cortisol level below 1.8 μg/dl excludes active CS at that time. However, Findling et al. [3] report a study with false-negative results in 18% of the patients with CD. Additionally, a potential misclassification of patients with increased CBG, acute and chronic illness or pseudo-CS is possible.

Recently, Papanicolaou et al. [4] published rates of 100% specificity and 93% sensitivity for the diagnosis of CS by using late-night salivary cortisol levels with a cut-off of 550 ng/dl. The probe is stable for 1 week which might represent an advantage for outpatient screenings. The promising results for this relatively simple test will have to be evaluated in the future.

Due to overall unsatisfactory rates of sensitivity and specificity of the first-line screening tests, some additional tests can be adopted to gain diagnostic precision.

Patients with CS often lack a normal circadian rhythm of cortisol secretion. This principle can be applied in measuring cortisol levels in hospitalized, sleeping patients at midnight. A cut-off of 1.8 μg/dl (or 7.5 μg/dl to increase specificity to 100%) has been adopted for highly sensitive results.

The classical 2-day low-dose dexamethasone test serves as another second-line screening test. Urine is collected on 2 baseline days and on the second day of dexamethasone administration. A decrease of free urinary cortisol less than 10 μg/24 h on the second day of dexamethasone is considered pathological. Alternatively, plasma cortisol can be measured at 9 h after the last dose of dexamethasone which should be less than 1.8 μg/dl (sensitivity and specificity more than 95%).

The 2-day low-dose dexamethasone test can be combined with CRH administration. Dexamethasone is given orally 0.5 mg every 6 h for 48 h and ovine-sequence CRH is administered (1 μg/kg i.v.) 2 h after the last dose of dexamethasone. If the plasma cortisol 15 min after CRH administration is above 1.4 μg/dl, CS is very likely. The test has been shown to be of value mainly to diagnose pseudo-Cushing states, since responses to CRH in pseudo-Cushing might be blunted due to chronic CRH secretion.

Differential Diagnosis of Cushing Syndrome

After CS is diagnosed, the differential diagnoses of CS have to be distinguished by measuring ACTH. ACTH-producing pituitary adenomas are the most likely cause of the ACTH-dependent CS. Other ACTH-dependent forms include ectopic ACTH or CRH secreting tumors. Alternatively, the glucocorticoid excess may be due to ACTH-independent adrenal pathologies like adrenal neoplasias or hyperplasia.

Hypercortisolism in combination with ACTH levels below 10 pg/ml at 9 h argues for an ACTH-independent CS and further imaging of the adrenals by CT or MRI scanning should be performed to reveal the type of adrenal pathology.

ACTH above >20 pg/ml suggests an ACTH-dependent cause of CS. Values in between 10 and 20 pg/ml can be further differentiated by performing a CRH test with measurement of plasma ACTH.

A battery of tests can help to distinguish the source of the ACTH overproduction.

The high-dose dexamethasone suppression (DST) test relies on the principle that ACTH secretion is suppressed in most corticotropic adenomas, at least partially, but not in ectopic tumors. Since this is not always true for some benign neuroendocrine tumors, the results of this test have to be interpreted cautiously.

Several versions of the high-dose DST exist (2-day oral high dose, 8 mg overnight oral, 4–7 mg i.v.). In all versions, plasma and urinary cortisol are evaluated before, during and after dexamethasone administration. A cut-off suppression of 50–80% is usually applied with a sensitivity of 60–80% to discriminate CD from ectopic tumors.

The CRH test and its interpretation are still under debate. This is due to the type of CRH used – for example i.v. 1 μg/kg or 100 μg synthetic ovine or human CRH–, biochemical parameters like ACTH increase 35–50% above baseline versus cortisol increase 14–20% and evaluated time points like ACTH measurement after 15–30 min or cortisol after 15–45 min.

New tests including substances like desmopressin, GH secretagogues and opiate agonists are still under investigation. Desmopressin for example might be an attractive agent for future research, because of its inexpensiveness, lack of severe side effects and its ability to increase ACTH secretion in 80–90% of patients with CD, but not in normal individuals.

After having performed the biochemical tests and an ACTH-dependent CS has been diagnosed, a gadolinium enhanced MRI should be performed. A lesion in the pituitary gland bigger than 6 mm in combination with the typical clinical symptoms and positive test results confirm the diagnosis of CD. However, it is important to bear in mind that in 10% of the normal population an incidentaloma, usually smaller than 5 mm, can be found. On the other hand, only in 60% of patients with CD a discrete pituitary adenoma will be detected.

In case of a negative MRI result, in patients with ACTH-dependent CS, bilateral inferior petrosal sinus sampling (BIPSS) with and without CRH test can be considered. Complications including deep vein thrombosis, pulmonary emboli and brain stem vascular damage rarely occur.

After the radiologist in a specialized center has catheterized both inferior petrosal sinuses (IPS), which is ideally verified by subtraction digital angiography, blood samples of ACTH are obtained basal, 3 and 5 min after CRH administration from both IPS and a peripheral vein. An IPS/peripheral ACTH ratio greater than 2 in the basal state or greater than 3 after CRH is consistent with CD and has a high sensitivity and specificity.

The diagnostic accuracy in identifying the site of a pituitary microadenoma is under discussion.

If BIPSS confirms the lack of a pituitary gradient, an ectopic source of ACTH production is likely and imaging techniques for neck, thorax and abdomen should be performed. Somatostatin analogue scintigraphy with ^{111}In-pentetreotide may be helpful in detecting occult somatostatin-receptor-positive tumors. PET is an interesting future option, but has not yet been established. Potentially adequate substances have been recently reviewed by Pacak et al. [5].

New Aspects in the Treatment of Cushing Disease

If diagnostic procedures reveal an ACTH-producing pituitary adenoma, transsphenoidal resection of the pituitary adenoma is the treatment of choice for the reduction of ACTH secretion. However, surgery is associated with significant post-operative morbidities. In cases of recurrence, pituitary irradiation or adrenalectomy is largely used as an adjunctive therapy.

Depending on the etiology, two different treatment modalities which still need to be validated are possible: reduction of pituitary ACTH production (cyproheptadine, sodium-valproate, bromocriptine or octreotide) or reduction of adrenocortical cortisol secretion (o,p'DDD, amino-gluthethimide, metyrapone and ketoconazole) (table 1) [2].

Lately, there was much progress in understanding the molecular mechanisms that control the function of the hypothalamic-pituitary-adrenal axis. Thus, new insights made it possible to identify potential drug targets for the treatment of CS caused by ACTH secreting pituitary adenomas. The role of dopamine agonists, somatostatin analogues and nuclear receptors such as retinoic acid receptors [retinoic acid receptor (RAR) and retinoid X receptor (RXR) and peroxisome proliferator-activated receptor-γ (PPAR-γ)] will be discussed.

Table 1. Medical treatment for Cushing syndrome

Substance	Action site	Mechanism
Cyproheptadine	central nervous system	serotonin antagonist
Sodium valproate	central nervous system	GABA-aminotransferase inhibitor
o,p'DDD	adrenal gland	glucocorticoid synthesis inhibitor
Amino-gluthethimide	adrenal gland	glucocorticoid synthesis inhibitor
Metyrapone	adrenal gland	glucocorticoid synthesis inhibitor
Ketoconazole	pituitary and adrenal gland	ACTH and glucocorticoid synthesis inhibitor
Bromocriptine	pituitary	dopamine agonist
Cabergoline	pituitary	dopamine agonist
Octreotide	pituitary	somatostatin analogue
SOM230	pituitary	multiligand somatostatin analogue
Retinoic acid	pituitary and adrenal gland	RAR-RXR ligand
Rosiglitazone	pituitary	PPAR-γ ligand

Dopamine Agonists

The significance of dopamine in the treatment of pituitary adenomas was first described in prolactinomas. Dopamine acts on the anterior and intermediate lobe of the pituitary gland to regulate PRL and MSH secretion [6, 7]. The identification of a functional dopamine D2 receptor (D2DR) in prolactinomas constituted dopamine agonists as a primary treatment for this type of pituitary tumors [8, 9].

Short-term treatment of patients with CD with the dopamine agonist bromocriptine decreased urinary/plasma cortisol levels in 42% of the cases and ACTH levels in 27% [10]. However, the efficacy of long-term treatment in bromocriptine-responsive patients was less than 50%.

Immunohistochemical analysis of 20 corticotropinomas revealed D2DR expression in 15 cases [11]. Treating corticotropinomas in primary cell culture with the new-generation dopamine agonist, cabergoline, decreased ACTH secretion in 60% of the cases [11]. The response to cabergoline correlated with D2DR expression and was abolished by the D2DR antagonist sulpiride, confirming that the treatment was specific for D2DR. Furthermore, as reported in the same study, cabergoline administration in patients with pituitary CD normalized cortisol levels in 40% of cases. All corticotropinomas from the cabergoline responders expressed D2DR. This study shows that cabergoline can be useful for the treatment of a subset of patients with CD.

Somatostatin Analogues

The general antisecretory action of somatostatin and its analogues prompted the testing of their efficacy for the treatment of CD. Although octreotide could effectively suppress ACTH secretion in patients with Nelsons' syndrome or ectopic CD, it was not effective in patients with pituitary dependent CD [10, 12, 13]. Octreotide was found to inhibit basal and CRH-induced ACTH secretion from corticotropinomas in primary cell culture [14]. However, this effect was abolished by hydrocorticosterone treatment, raising the hypothesis that patients with CD cannot respond to octreotide treatment due to high cortisol levels. It was recently shown that dexamethasone treatment decreases the mRNA expression of somatostatin receptor type (SSTR) 2, which is the main mediator of octreotide action [15]. Most corticotropinomas were found to predominantly express SSTR5, indicating that somatostatin analogues with high affinity to SSTR5 could be useful for Cushing treatment [15, 16]. SOM230 is a multiligand somatostatin analogue with high affinity to SSTR1, 2, 3 and 5. Compared to octreotide, SOM230 has 2.5 times less affinity for SSTR2, but 40 times more affinity for SSTR5 [17]. Treatment with SOM230 for 72 h decreased ACTH secretion in 3 of 5 human corticotropic adenomas in primary cell culture, while octreotide did so in only one case [15]. SOM230 was found to be more efficient than octreotide in reducing ACTH secretion in rodents, although its efficacy in patients with CD remains to be determined [18].

Retinoic Acid

Retinoids, comprising the native and synthetic derivatives of vitamin A, are promising agents for the prevention and treatment of human cancers including breast and lung cancer. The biological effects of retinoids are mainly mediated by their nuclear receptors, RAR and RXR [19, 20].

Using AtT-20 pituitary corticotropic tumor cells, we have recently demonstrated that retinoic acid inhibits ACTH secretion in vitro by inhibiting the transcriptional activity of the transcription factors AP1 and Nur on the POMC gene, which encodes ACTH [21]. However, this inhibitory action of retinoic acid seems to be restricted to ACTH-secreting tumor cells, since normal ACTH-secreting cells and other normal pituitary cells are not affected by the treatment, demonstrating a specific effect on tumor cells probably related to the distinct differentiation degree of normal cells versus tumor cells. Treatment of human corticotropinomas in primary culture with retinoic acid resulted in the inhibition of the ACTH production by these cells. We have also shown that retinoic acid inhibits cell proliferation and induces apoptosis in ACTH-secreting tumor cells [21].

When athymic nude mice were inoculated with corticotropic tumor cells, control animals develop large subcutaneous corticotropic tumors, whereas no tumors were observed in mice injected with retinoic acid-treated cells, indicating that the antiproliferative effects of retinoic acid are also effective in vivo. Moreover, administration of retinoic acid to mice that already had experimental ACTH-secreting tumors, resulted in the inhibition of tumor growth. Plasma levels of ACTH and cortisol were reduced in retinoic acid treated mice, compared with controls. This inhibition resulted in the reversion of adrenal hyperplasia and skin atrophy, both characteristic symptoms of CS [21].

In an ongoing study with dogs suffering from CD, preliminary data shows that the treatment with retinoic acid decreased ACTH and cortisol levels and improved clinical symptoms in respect to control animals treated with ketoconazole [Castillo et al., unpubl. data].

Retinoic acid treatment might represent a potential therapeutic option to inhibit ACTH and cortisol production, as well as tumor growth in CD.

PPAR-γ Ligands

PPAR-γ heterodimerizes with RXR to function as a transcription factor that binds to specific response elements called peroxisome-proliferator response elements. In normal human pituitary tissue, PPAR-γ expression is restricted to ACTH-secreting cells [22]. By contrast, PPAR-γ is highly expressed by all pituitary tumor cell subtypes [23]. Treatment of murine and human corticotropinomas with the synthetic PPAR-γ ligand rosiglitazone induced cell-cycle arrest and apoptosis and suppressed ACTH secretion. In vivo, rosiglitazone prevents the development of corticotropic tumors in nude mice. Levels of plasma ACTH and serum corticosterone were reduced in rosiglitazone-treated mice, compared with controls [22].

The administration of rosiglitazone in patients with CD did not give promising results. Eight of 14 patients treated with rosiglitazone did not respond to the treatment. Six of 14 patients responded after 2 months of treatment by lowering ACTH and serum cortisol and normalization of free urinary cortisol. Of these 6 patients, 2 were followed for 7 months and showed a mild clinical improvement [24]. Additional studies are needed before conclusions can be made.

Since a combination of retinoic acid and rosiglitazone might allow the reduction of pharmacological dosages of both drugs to be administered therapeutically, we investigated whether the RAR and RXR might have synergistic inhibitory actions with PPAR-γ ligands. In murine corticotropic tumor cells, the combination of rosiglitazone and retinoic acid led to a synergistic inhibition of ACTH secretion [unpubl. data]. Thus, combined treatment with retinoic acid and rosiglitazone may increase the efficiency of the treatment leading to a synergistic antiproliferative and antisecretory effect on tumor cells.

Conclusion

If CS is suspected, it is recommended to measure cortisol in more than one 24-hour urinary collection and to perform a low-dose dexamethasone suppression test as the first-line screening test according to the diagnostic algorithm of the consensus statement. However, false-positive results are common. Late night salivary cortisol might be an interesting future option, but more published data are needed.

The diagnostic evaluation should not proceed until the diagnosis of CS is firm.

ACTH levels, CRH stimulation tests and the high-dose DST as well as CT/MRI scanning are the most useful and common tests for the differential diagnosis of CS.

BIPSS is recommended for patients with ACTH-dependent CS whose clinical, biochemical and radiological studies are discordant.

At present, there is no effective pharmacological therapy that has been clinically tested to control ACTH oversecretion by pituitary tumors. New studies on dopamine agonists and somatostatin analogues such as cabergoline and SOM230, respectively, pointed the potential importance of these drugs in the treatment of CD.

Recent advances in elucidating the function of nuclear receptors have resulted in the development of novel approaches. PPAR-γ ligands and retinoic acid might represent potential therapeutic options to inhibit ACTH production, as well as tumor growth in CD. The efficacy of these treatments in patients with CS still needs to be tested in clinical trials.

References

1 Arnaldi G, et al: Diagnosis and complications of Cushing's syndrome: a consensus statement. J Clin Endocrinol Metab 2003;88:5593–5602.
2 Labeur M, et al: New perspectives in the treatment of Cushing's syndrome. Curr Drug Targets Immune Endocr Metabol Disord 2004;4:335–342.
3 Findling JW, Raff H, Aron DC: The low-dose dexamethasone suppression test: a reevaluation in patients with Cushing's syndrome. J Clin Endocrinol Metab 2004;89:1222–1226.
4 Papanicolaou DA, et al: Nighttime salivary cortisol: a useful test for the diagnosis of Cushing's syndrome. J Clin Endocrinol Metab 2002;87:4515–4521.
5 Pacak K, Eisenhofer G, Goldstein DS: Functional imaging of endocrine tumors: role of positron emission tomography. Endocr Rev 2004;25:568–580.
6 Ben-Jonathan N: Dopamine: a prolactin-inhibiting hormone. Endocr Rev 1985;6:564–589.
7 Stack J, Surprenant A: Dopamine actions on calcium currents, potassium currents and hormone release in rat melanotrophs. J Physiol 1991;439:37–58.
8 De Camilli P, Macconi D, Spada A: Dopamine inhibits adenylate cyclase in human prolactin-secreting pituitary adenomas. Nature 1979;278:252–254.
9 Colao A, Annunziato L, Lombardi G: Treatment of prolactinomas. Ann Med 1998;30:452–459.
10 Miller JW, Crapo L: The medical treatment of Cushing's syndrome. Endocr Rev 1993;14: 443–458.
11 Pivonello R, et al: Dopamine receptor expression and function in corticotroph pituitary tumors. J Clin Endocrinol Metab 2004;89:2452–2462.
12 Lamberts SW, Uitterlinden P, Klijn JM: The effect of the long-acting somatostatin analogue SMS 201–995 on ACTH secretion in Nelson's syndrome and Cushing's disease. Acta Endocrinol (Copenh) 1989;120:760–766.

13 von Werder K, Muller OA, Stalla GK: Somatostatin analogs in ectopic corticotropin production. Metabolism 1996;45(suppl 1):129–131.
14 Stalla GK, et al: Octreotide exerts different effects in vivo and in vitro in Cushing's disease. Eur J Endocrinol 1994;130:125–131.
15 Hofland LJ, et al: The multi-ligand somatostatin analogue SOM230 inhibits ACTH secretion by cultured human corticotroph adenomas via somatostatin receptor type 5. Eur J Endocrinol 2005;152:645–654.
16 Nielsen S, et al: Expression of somatostatin receptors on human pituitary adenomas in vivo and ex vivo. J Endocrinol Invest 2001;24:430–437.
17 Bruns C, et al: SOM230: a novel somatostatin peptidomimetic with broad somatotropin release inhibiting factor (SRIF) receptor binding and a unique antisecretory profile. Eur J Endocrinol 2002;146:707–716.
18 Silva AP, et al: Regulation of CRH-induced secretion of ACTH and corticosterone by SOM230 in rats. Eur J Endocrinol 2005;153:R7–R10.
19 Chambon P: A decade of molecular biology of retinoic acid receptors. FASEB J 1996;10:940–954.
20 Zhang XK, et al: Retinoid × receptor is an auxiliary protein for thyroid hormone and retinoic acid receptors. Nature 1992;355:441–446.
21 Paez-Pereda M, et al: Retinoic acid prevents experimental Cushing syndrome. J Clin Invest 2001;108:1123–1131.
22 Heaney AP, et al: Functional PPAR-gamma receptor is a novel therapeutic target for ACTH-secreting pituitary adenomas. Nat Med 2002;8:1281–1287.
23 Heaney AP, Fernando M, Melmed S: PPAR-gamma receptor ligands: novel therapy for pituitary adenomas. J Clin Invest 2003;111:1381–1388.
24 Ambrosi B, et al: Effects of chronic administration of PPAR-gamma ligand rosiglitazone in Cushing's disease. Eur J Endocrinol 2004;151:173–178.

Prof. Dr. med. Günter K. Stalla
Internal Medicine, Endocrinology and Clinical Chemistry
Max-Planck-Institute of Psychiatry
Kraepelinstrasse 10
80804 Munich (Germany)
Tel. +49 893 0622 270, Fax +49 893 0622 460, E-Mail stalla@mpipsykl.mpg.de

Author Index

Subject Index

Pituitary today.

DATE			